EUREKA!
Physics of Particles, Matter and the Universe

EUREKA!
Physics of Particles, Matter and the Universe

Roger J Blin-Stoyle, FRS
Emeritus Professor of Physics
University of Sussex

Institute of Physics Publishing
Bristol and Philadelphia

British Library Cataloguing-in-Publication Data

A catalogue record for this book is available from the British Library.
ISBN 0 7503 0415 4 (hbk)
ISBN 0 7503 0416 2 (pbk)

Library of Congress Cataloguing-in-Publication Data

Blin-Stoyle, R. J. (Roger John)
    Eureka! : physics of particles, matter and the universe / Roger Blin-Stoyle
        p. cm.
    Includes index.
    ISBN 0-7503-0415-4 (hc : alk. paper).—ISBN 0-7503-0416-2 (pbk.
  : alk. paper)
        1. Physics. I. Title.
    QC21.2.B567 1997
    530—dc21   97-19999
                                                                     CIP

Consultant Editor: Frank Close, FRS

Published by Institute of Physics Publishing,wholly owned by The Institute of Physics, London

Institute of Physics Publishing, Dirac House, Temple Back, Bristol BS1 6BE, UK

US Editorial Office: Institute of Physics Publishing, The Public Ledger Building, Suite 1035, 150 South Independence Mall West, Philadelphia, PA 19106, USA

Typeset by Mackreth Media Services, Hemel Hempstead, Herts
Printed in the UK by J W Arrowsmith Ltd, Bristol

To Helena and Anthony

CONTENTS

PREFACE

There is a general perception that physics is a difficult science to understand. This arises for two main reasons. First, it is the most quantitative of all the sciences and, inevitably, the detailed description of its underlying theories is mostly couched in very advanced mathematical language. Second, at the most fundamental level, it deals with processes and phenomena on time and space scales inconceivably smaller or larger than those experienced in our everyday life. In other words it deals with a great deal of alien territory in terms of, for many people, an alien language. Hence the aforementioned 'difficulty'.

This is not to say that what physics has achieved and is trying to achieve cannot be communicated to the lay person. At one extreme this can be done by attempting entirely qualitative descriptions and explanations of physical phenomena. A great many words are used in the process but some idea of what physics is about can be conveyed. A closer approach to the real nature of physics is to deal with physical processes just a little more quantitatively, occasionally using the sort of elementary mathematics met with regularly by young secondary- or high-school pupils. This is the approach adopted in this book, which attempts to give a brief, matter-of-fact, account of what the whole of physics is about at all levels of scale—from the ultimate constituents of matter, through nuclei, atoms and molecules, to the behaviour of the different forms of matter and, finally, on to stars, galaxies and the nature of the universe itself.

It is a short book requiring no previous detailed knowledge of physics other than a general awareness of everyday physical concepts such as matter, force, energy, speed, space and time. It starts with down-to-earth physical processes including topics that are key parts of the National Curriculum in the UK. Parts of this could, no doubt, be omitted by some readers—but revision of some early learning is not, perhaps, a bad thing! The book then moves into less familiar but more exciting and challenging territory. The hope is that it will illuminate the nature of the whole of physics for a wide variety of readers—school pupils, college or university students, teachers at all levels and any lay person who wishes to know about physics and is prepared to countenance the occasional algebraic symbol!

A Glossary is provided which, first of all, gives a brief account of the way in which very small and very large numbers are represented and it is suggested that the uninitiated should study this section carefully before embarking on the main text. It then goes on to list the units which are used to measure physical quantities and also gives the values of some of the key physical constants (e.g. the speed of light). Finally, brief definitions are given of physical terms which are used in the text.

In concluding this Preface I would like to thank all those with whom I have discussed physics over the last 50 years—school pupils, teachers, undergraduates, research students, fellow researchers and colleagues. All have contributed in their very different ways to whatever understanding I have managed to communicate in this book and to the enjoyment of my career as a physicist.

Roger Blin-Stoyle
May 1997

CHAPTER 1

UNDERSTANDING THE WORLD AROUND US

Towards a Theory of Everything?

1.1 What is Physics?

Physics is that branch of science which seeks to understand the behaviour and properties of matter at all levels of scale. At one extreme it is concerned with the fundamental constituents of matter—the so-called elementary particles—and with atoms and molecules. The latter are the building blocks of everyday matter and it is in terms of them that it interprets the very varied properties of solids, liquids and gases. On the larger terrestial scale it studies the behaviour of the air and ocean masses, climate and the environment. Finally, at the other extreme it concerns itself with the structure of stars and stellar systems and, ultimately, the nature and evolution of the universe itself.

Such a description implies that physics encompasses most of science. It is certainly true that physics underlies and underpins most, if not all, scientific understanding; however, as science developed over the centuries, many areas have come to be regarded and organized as separate, although related, sciences. Thus the interactions between and processes involving simple or complex molecular structures are generally classified as *chemistry*, whilst the study of living matter with all its extreme molecular complexity is classified as *biology*. However, the dividing lines are extremely fuzzy and are spanned by various 'bridging' sciences such as *chemical physics*, *biophysics* and *biochemistry*. Further, physics concerned with larger-scale phenomena is generally referred to by other names. Thus, at the terrestrial level, we have *meteorology*; at the stellar level we have *astronomy* and *astrophysics*; and, at the scale of the whole universe, we have *cosmology*.

The primary thrust of physics is the intellectual satisfaction of achieving understanding of a wide variety of phenomena, but, beyond this, such understanding enables the production of materials, devices, structures and processes which can be of immense benefit (although not always!) to mankind. Most modern technology—transport, communications, electronic wizardry in the home, commerce and industry, medical diagnostics and therapies . . . —are based on advances in physical understanding. The efforts of many physicists—*applied physicists, medical physicists, material scientists,*—are dedicated to developing applications of this kind. And, of course, the whole of engineering—electronic, electrical, mechanical and even civil— depends to a greater or lesser extent on physical processes and the physical properties of matter.

1.2 The Nature of Understanding

Understanding can have many facets and can be achieved at varying depths. As far as physics is concerned, preliminary understanding is obtained when a group of similar phenomena can be explained in terms of some overall basic idea. For example, the orbits of the different planets about the sun can be understood in considerable detail in terms of their motion under the gravitational attraction of the sun. The 'basic idea' involves the general specification of the way in which bodies of different mass move in space when subject to an external force and the specification of the nature of the force of gravity between two massive bodies, in this case the sun and the planet.

Such a 'basic idea' is called a *theory* and, in physics, a theory is invariably specified in mathematical form. It then enables quantitative relationships to be derived between the various measured quantities of the phenomena under study and, if the theory is successful, these relationships should agree with those observed. Thus, for the example quoted, knowledge of the position, speed and direction of motion of a planet at a given time can be used to predict these same quantities at any subsequent time. The extent to which such predictions are born out by experiment is a measure of the success of the theory.

In general, then, a theory is postulated to account for a set of experimental data. It also enables the prediction of other previously unmeasured data and its correctness must be judged by whether its predictions are confirmed by further new experimental observations. If they are—well and good, and further checks are made. If they are not—the theory has to be modified or even radically changed and further experimental tests carried out. So, gradually, successful theories and deeper understanding emerge through the continual cycle

experiment → theory → test predictions → revise theory →
test predictions → revise theory → and so on.

However, it must be recognized that theories cannot be proved absolutely—that would require an infinite number of tests—and all theories must be regarded as provisional. You never know whether some new data will unseat it. On the other hand, if every test agrees with the theory, then there is increasing confidence that the theory is correct. In some cases confidence in the theory is so great that its key features have been referred to as 'laws'; for example Newton's laws of motion, the laws of thermodynamics and what are known as conservation laws (*q.v.*).

As physics has progressed over the years, theories about the behaviour of matter have been continually developed. For example, by the beginning of the 19th century there were crude theories about the electrical and magnetic properties of matter which gave understanding of such phenomena as frictional electricity (polish an inflated balloon and it will pick up small pieces of tissue paper) and the forces between magnets. Then, in the early decades of the 19th century, Oersted discovered that a wire carrying an electric current behaved like a magnet and Faraday demonstrated that moving a magnet near a wire produced an electric current; electricity and magnetism are clearly related to each other. Considerable understanding had also been achieved about the behaviour of light—for example, how it passed through lenses, the formation of rainbows and its speed of travel. Finally, by the end of the 19th century, it became clear that electricity, magnetism and light could *all* be understood in terms of a single all-embracing theory—known as electromagnetic theory—formulated by James Clerk Maxwell.

This advance is a spectacular example of the way in which physical understanding progresses. Gradually more and more phenomena are being understood in terms of fewer and fewer basic theories. Eventually it may be that the end of the road will be reached in which a core of fundamental ideas are incorporated into a comprehensive theory able to account for all physical phenomena—a 'theory of everything'. To find such a theory is, perhaps the ultimate goal of physics. Further, however simple such a final theory might be, there will be no escaping the fact that most physical phenomena will still be extremely complicated. Discussion of this sort of issue and the consideration of whether such a final theory reveals in some sense the 'mind of God' has occupied many pages in recent books.

1.3 The Problem of Complexity

Some phenomena in physics have an apparent simplicity in their make-up. For example, the motion of a planet about the sun involves just two bodies—the sun and the planet—and the specification of the gravitational force between them. With such a simple system it is possible to calculate, with essentially as much accuracy as is desired, the precise details of the planetary motion. Slightly more complicated systems involving just a few basic entities, whether they be planets, simple atoms or molecules, can also be treated with reasonable accuracy so that agreement between theory and experiment can be checked in considerable detail. Even with more complex systems containing up to a few hundred components, for example large atoms or atomic nuclei, it is possible to construct (see section 1.4) reasonably quantitative and testable theories of their behaviour.

However, most physical systems have a *very* large number of components. For example, in a pin head there are around 100 million million million (10^{20}) atoms of iron; a litre of air contains around 100,000 million million million (10^{23}) molecules. These numbers are, of course, as nothing compared with the number of atoms or molecules in the oceans or atmosphere or, at an even more extreme level, in a star such as the sun.

In general, even if the nature of the forces between atoms and molecules were fully understood—and a lot is now known about them—it would be quite impossible to work out the detailed motion of every atom and molecule in such *macroscopic* systems. (Here, *macroscopic* means large enough to be observed by the naked eye, as distinct from *microscopic*.) However this is not to say that no progress can be made in understanding the behaviour of such complex systems.

Planets are very large macroscopic systems, yet, as has already been mentioned, their motion around the sun can be understood in great detail. Here, the understanding that has been achieved is about the motion of this macroscopic system *as a whole*; it is not about the details of the motion of the individual and virtually innumerable atoms and molecules which constitute the planet. What we do know about their motion is that *on average* they are together moving in a very well defined orbit about the sun.

Other average or macroscopic properties of matter can be similarly understood, for example, the pressure exerted by a gas on its container, the conduction of heat or electricity through a metal, the freezing of a liquid when it is cooled or its vaporization when it is heated. In all such examples, and many more could be quoted, the understanding achieved is in terms of the average behaviour of the component atoms or molecules. The approach to this form of understanding is a *statistical* one and this is possible and meaningful simply because of the very large number of atoms or molecules involved. Some macroscopic physical systems are still essentially simple in their structure. For example, in a pure substance there is only one sort of atom or molecule to consider and, sometimes, they may be arranged in an extremely tidy and symmetrical way. This occurs in *crystalline* substances where, in the simplest case, the atoms are arranged in straight rows and columns and are simply located at the the corners of a cubical lattice. As we shall see, they will be vibrating about their average positions but, because they are essentially localized, it becomes possible to make relatively simple theories about their individual motions. For such systems full and deep understanding of their physical properties is frequently obtained.

However, many entities or systems are far more complicated. They may not only have a vast number of component atoms and molecules, but also many different varieties and the overall structure can be unimaginably complicated. Examples of such systems are the different types of biological material, the human brain and, on the larger scale, weather systems. Here, although understanding of some general features can be obtained in terms of the behaviour of component parts, it generally proves impossible to give a detailed account of their behaviour; they are just too complicated. Weather forecasting is a well known example. Short-term (of the order of a few hours, up to a day) forecasts are usually reasonably accurate but longer-term forecasting is notoriously inaccurate. The problem is that the evolution of such systems over time is a very complicated process and, further, depends extremely sensitively on the very fine details of the intial state of the system. In the case of the weather, the example often quoted that the development of the weather in the USA can be affected significantly by the beating of a butterfly's wings in South America some weeks before is probably an exaggeration, but nevertheless indicates the nature of the problem. With such systems, however well the nature of their microscopic components and the way they interact with each other is understood, and even if the underlying theory is completely deterministic (events in the system are fully determined by preceding events), the sheer complexity of the systems means that prediction of their detailed behaviour cannot be achieved. This unpredictability and the systems which exhibit it are encompassed in an area of physics known as *chaos* or *chaology* which, over the last few years, has been receiving a great deal of attention.

In summary, phenomena in the physical world range over those which can be described in terms of the behaviour of a few basic entities (fundamental particles, atoms, molecules, planets, stars) which, generally, have a sub-structure but whose details are irrelevant to the phenomenon being considered, through to those which can only be described in terms of large numbers and varieties of entities. When there are relatively few it is generally possible to construct theories which enable understanding and predictions of detailed behaviour to be made, but as the number of entities involved in a phenomenon increases only broad and general behaviour can be understood. Detailed understanding becomes less and less possible and eventually, with the most complicated systems, behaviour can become virtually unpredictable and chaotic.

1.4 Conceptual Models in Physical Theory

Although the preceding discussion implies that in general it is difficult to deal in fine detail with theories of systems involving more than a few basic entities, it has been possible to devise simple approximate theories or conceptual models which do enable significant understanding of some properties of such systems to be achieved. It has already been mentioned that for the purposes of understanding the orbital motion of a planet, the complexity of its internal constitution can be ignored: it is sufficient to treat the planet as though all of its mass were concentrated at a point (known as its *centre of mass*). Similarly to understand the way in which a liquid flows (the science of *hydrodynamics*) it is generally sufficient to treat the liquid as a continuum (i.e. absolutely uniform throughout) and to ignore its atomic/molecular structure. In each of these cases, what are in reality extremely complex structures are represented by simple conceptual models which enable understanding of certain aspects of their behaviour to be well understood. Of course, in the former case, if we wished to understand geological behaviour such as earthquakes or the eruption of volcanoes, the model would be useless, as would be the liquid model if we wished to understand freezing and vaporization.

Models are used extensively in physics at both the macroscopic and microscopic levels to enable understanding of a limited range of features. The more features that can be understood in terms of the model the better it is, and refinement of many standard models of physical behaviour are always being sought. Care has to be taken however not to confuse a model with reality. A model is just a simple and manageable representation of that reality which enables some of its physical properties to be understood. In physics, such models are usually mathematical in nature.

1.5 Human Experience of the Physical World

On embarking on this journey through the world of physical phenomena it is extremely important to recognize that our own direct experience of the physical world is miniscule. Consider our spatial experience. Being generous we probably have a feeling for something as small as 0.1 mm (10^{-4} m) and as large as the earth which has a diameter of about 12,000 km (roughly 10^7 m), but, as we shall see, many physical phenomena involving the fundamental constituents of matter take place within distances of around 10^{-15} m. At the other extreme the visible universe extends to a distance of around 10^{26} m. Similarly, with continuing generosity, our feeling for time may extend down to 1/1000 of a second (10^{-3} s) through to, if we are lucky, 100 years (about 10^9 s). These figures are to be compared with the time scale of fundamental particle processes which can be as short as 10^{-23} s and the age of the universe which is around 15,000 million years (about 10^{17} s). This is not to mention the extreme conditions which occurred in the big bang, when the universe came into being as a result of a gigantic explosion and when formidable changes took place in infinitesimal time intervals

This comparison of human experience in space and time with that of physical phenomena is shown diagramatically in Figure 1.1. Because of the paucity of our experience in the very small and very large realms of space and time it should come as no surprise if physical processes take place which are completely at variance with our very limited everyday experience and expectations. We shall come across some very strange phenomena which it will be hard to accept as 'natural'. To anticipate just one, we shall find that when talking about the basic constituents of matter (the elementary particles) it becomes impossible to say anything about such a particle's state of motion if we know *precisely* where it is! This is completely contrary to our experience of a ball on a billiard table, where we can know where it is and how it is moving.

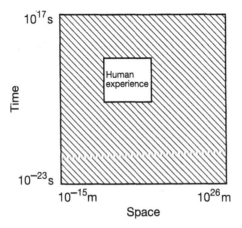

Figure 1.1: *Human experience of space and time in the physical world.*

There is also another aspect of our experience which is very limited, namely our direct experience of the dimensions of space and the flow of time. We are all completely conscious of three dimensions of space: up (or down), sideways (left or right) and forward (or backward). We are also conscious of the passage of time and so, if we want to specify the location of an event in our lives, we do so in terms of the position in space (specified in terms of the position of the event in the three dimensions just referred to) and the time at which it happens. Space and time are treated by us as completely separate. However, when we come to discuss relativity in Chapter 7 we shall find that space and time are intimately related and that our natural perceptions of their separateness have to be abandoned when dealing with fast-moving highly energetic objects.

Further, in considering the nature and behaviour of the fundamental constituents of matter, we shall learn that current theories imply that there may be many more dimensions to consider than the four (three space and one time) evidenced by our own daily experience. Such a suggestion is hard, if not impossible, for us to accept. However, put yourself in the position of an imaginary being only experiencing two spatial dimensions—

forward (backwards) and sideways. Imagine this being living on the surface of a sphere and only being conscious of motion in these two directions on this surface. Such a being, with its limited experience, would find it impossible to conceive of an upward (downward) dimension and would believe that its universe was unbounded—i.e. there was no edge to it. In other words, its universe would appear to be infinite. *We,* on the other hand, know that it is finite and simply the surface of a sphere. With this example we, recognizing our limited three-spatial-dimensional experience, should perhaps not be surprised if more dimensions are needed to give a full description of the physical world. We should also recognize that although our universe appears to be unbounded it may not, in fact, be infinite in extent.

So this chapter ends with the warning that as we progress to considering physical phenomena on the very small or very large scale and also at very high energies—all outside our own direct experience—then 'common sense' derived from that experience will not necessarily be a good guide to achieving understanding.

1.6 Moving Forward

Having indicated briefly the coverage of physics, the nature of physical understanding and the dangers of using our own experience as a guide to this understanding, let us now consider some aspects of the physical world which *do*, in fact, relate easily to our everyday experience.

CHAPTER 2

EVERYDAY EXPERIENCE OF MOTION AND ENERGY

Forces and their Effects

2.1 Motion and Forces

There is clearly a great deal of very varied motion in the world around us. Even those entities which appear to be stationary—for example the items of furniture in our rooms and the objects in and on them—are moving at high speed as the earth rotates and moves around the sun. Further, at the other extreme, the atoms and molecules from which they are constituted are, as we shall see, in incessant motion. It is therefore essential to understand at an early stage the nature of motion and how it can be changed.

First, to state the virtually obvious, the motion of a body is changed when a force is exerted on it where a force is characterized by two features—its *magnitude* and its *direction*. (Here it should be noted that entities specified by these two characteristics are known as *vectors*.) By 'changed' is meant that the body speeds up (accelerates), slows down (decelerates) and/or changes the direction in which it is travelling. To start a supermarket trolley moving (i.e. to change its motion—from rest to moving) it has to be pushed; the pusher exerts a force on it. Similarly a force has to be applied to turn it round a corner. Of course, to keep the trolley moving at a steady speed in a straight line a continual push is still required and yet the motion is not changing. Here it must be realized that there is another force influencing the motion, namely *friction*, and in steady motion the 'push' and 'frictional' forces just balance. In other words the net force on the trolley is, in fact, zero and hence its motion does not change. If there were no friction then no push would be required to keep the trolley moving steadily. This state of affairs is, for example, nearly reached when an object such as an ice puck, experiencing very little friction, slides over ice. The statement that a body's motion only changes when a force is exerted on it was formally enunciated by Isaac Newton in the 17th century and is incorporated in his *First Law of Motion*.

Newton's First Law of Motion. *A body continues in its state of rest, or uniform motion in a straight line, unless acted upon by an external force.*

In a moment we will consider in a little more detail how the change in motion brought about by a force is related to its strength and direction, but before doing that we should consider the nature of force. Everyone is familiar with the force exerted on an object when it is pushed or pulled. Such a force is generally transmitted by direct contact between the pusher or puller and the object experiencing the force. But the force may also be transmitted through an intermediate agency—pushing a stone with a stick, hitting a ball with a racket, controlling a kite with a cord etc.

Familiar to most will also be forces which are transmitted without any material contact, for example, the force exerted by a magnet on a piece of iron. Place a magnet near some iron filings and they will jump and attach themselves to it; wave a magnet near a compass needle and the needle will move. Here it will be recognized that magnetic forces can be repulsive as well as attractive; put two compass needles close to each other and the two north-seeking poles will move apart from each other. The iron filings and the compass needle change their state of motion under the influence of the magnet and a force, known as a magnetic force, is being exerted on them. Similarly, if a balloon is rubbed against a piece of material it can pick up pieces of tissue paper. In this case an electric force is coming in to play. It is the same type of force which raises the hairs on a hand or arm when placed close to a television screen. Finally, in this context, there is the force of gravity which pulls a ball down to the ground when thrown into the air and which keeps objects—including ourselves—firmly on the face of the earth and keeps planets orbiting about the sun. Gravity, unlike magnetic and electric forces, is a force which is *always* attractive; it attracts the moon to the earth, the earth to the sun and is, in fact, experienced between *all* material objects. It is very weak, however, and is only noticeable when at least one of the objects is *very* massive (e.g. the earth).

Magnetic, electric and gravitational forces, which are fundamental to understanding the behaviour and properties of matter at all levels of scale, are effective between bodies without there being any obvious direct physical contact between them. With such forces, the closer the two bodies experiencing them are, the stronger the force; you will not be able to detect the influence of a magnet on a compass needle placed on the other side of a room. The magnetic force dies away slowly as the distance from the magnet increases; similarly with electric and gravitational forces. The objects exerting such forces are surrounded by a 'field of influence' producing what might be called a 'stress' in space which becomes weaker the further you are away from the objects. It is conventional to refer to them as *magnetic, electric* and *gravitational fields*. In due course (Chapters 8 and 9) it will be explained how such 'action at a distance' forces and fields are propagated but, for the moment, just accept that they exist.

2.2 Force, Mass and Acceleration

In the previous section it was recognized that to move a trolley from rest required the application of a force. To move a car from rest would require a much greater force—a car has much greater resistance to motion or inertia. A measure of this inertia is what is called the *mass* of the trolley or car. Mass is intimately related to *weight* but is fundamentally different. The weight of an object is the gravitational force exerted on it by the earth and is measured, for example, by weighing it using a spring balance. An object weighed on the moon will have one-sixth of its weight on the earth simply because the moon is smaller and less massive than the earth and therefore exerts less gravitational attraction. Mass, on the other hand, is intrinsic to the body and has the same value wherever the body is; it is essentially just as hard to move a car from rest on the moon as it is on the earth!

The effect of exerting a force on a body is to make it move faster in the direction of the force; the body *accelerates*. If this is the only force acting on the body then the acceleration will be steady and the body will move faster and faster. The size of this acceleration is proportional to the size of the force and, as should be expected from our discussion of the trolley and the car, will be smaller the more massive the body. In fact the relationship between force, mass and acceleration is very simple

$$\text{acceleration} \ = \ \frac{\text{force}}{\text{mass}} \ .$$

This relationship is enshrined in *Newton's Second Law of Motion*.

Newton's Second Law of Motion. *The acceleration of a body is proportional to the force applied and is in the direction of that force.*

In the above equation we see that the constant of proportionality is the inverse of the mass of the body. Of course, if a body is already in steady motion and a force is applied in a direction *opposite* to that motion then *deceleration* proportional to the force takes place. The strength of a force is measured in what are called *newtons* (denoted by N) where one newton (1 N) is the force needed to give one kilogram (1 kg) an acceleration of one metre per second per second (1 m s^{-2}).

Newton also formulated a *Third Law*.

Newton's Third Law of Motion. *When two bodies interact with each other the force on the first body due to the second is equal and opposite to the force on the second body due to the first.*

For example, a weight placed on a table exerts a downward force on the table due to the pull of gravity. In turn the table exerts an equal upward force on the weight (see figure 2.1). If this (reaction) force were bigger there would be a net upward force

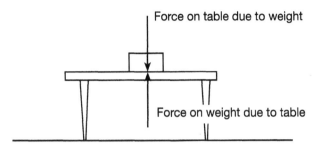

Figure 2.1: *Equal and opposite forces acting on a weight and a table.*

and the weight would move upwards; similarly if less than the downward force the weight would move downwards. Obviously neither of these situations can occur!

It is interesting at this stage to say a little more about the force of gravity. This force is proportional to the product of the masses of the two bodies interacting. In mathematical terms, if the masses of the two bodies (measured in kilograms) are m and M and their distance apart (measured in metres) is r, then the magnitude of the force F (measured in newtons) which each experiences pulling it towards the other is given by

$$F = \frac{GmM}{r^2}$$

where G is known as the *gravitational constant* and has the value $G = 6.67 \times 10^{-11}\,\mathrm{N\,m^2\,kg^{-1}}$. It is a measure of the strength of the gravitational interaction Here it is important to note that we have been discussing 'mass' in two different ways. As first introduced it is that quantity which specifies the *inertia* of a body and which determines the degree to which the body accelerates when a force is applied. In this context it is referred to as the *inertial mass*. Its second use has been as the quantity which determines the size of the gravitational force a body experiences due to another body as given in the above formula. In this second context it is referred to as the *gravitational mass*. The important point to note is that we find that these two different masses are *identical*.

The above law of gravitational attraction means that, on the earth, the gravitational forces experienced by different bodies are simply proportional to their mass since the mass of the earth is obviously common to all situations. Since the acceleration produced by this force is *inversely* proportional to the mass it follows that the acceleration down to the earth of a body dropped from a height is the same whatever its mass; the more massive the body the stronger the force of gravity, but the harder it is to accelerate it. This is not quite observed in practice since there is friction from the air (air resistance) and so in reality the net force on a falling body is gravity less air resistance and the latter will be different for differently shaped bodies and for different speeds of fall. However, if the bodies are reasonably heavy so that air resistance is negligible compared with the gravitational force then they will fall with very nearly the same acceleration. This was established first by Galileo in the 16th century when he is believed to have demonstrated this by dropping objects from the leaning tower of Pisa. On the moon, where there is no atmosphere, a lead weight and a feather will fall at the same speed. It is interesting to note that if the falling object on the earth is, for example, a person wearing a parachute, then the air resistance increases significantly as the speed of fall increases until, quite soon, it is equal to the force of gravity but is, of course, in the opposite direction. There is then zero net force and therefore zero acceleration so the falling person no longer accelerates and travels down to the earth at a constant and reasonably safe speed. Without a parachute the air resistance is much less and only balances the gravitational force at a much higher speed.

2.3 Momentum and Angular Momentum

There is another concept which is very useful in discussing motion, namely *momentum*. We are all familiar with the qualitative idea of momentum; a body with high momentum, for example a moving car or a bullet in flight, requires a large force to bring it to rest or, to put it another way, the moving body exerts a large force on whatever is stopping it. Momentum is clearly related to the mass of the body, its speed and its direction of motion. Its magnitude is, in fact, simply the product of the mass of a body and its velocity, where by velocity we mean the speed of the body and also its direction of motion. Denoting the magnitude of momentum by p, mass by m and velocity by v we can therefore write

$$p = mv.$$

Velocity is another *vector* quantity since it has magnitude and direction and so, therefore, is momentum—a bullet with high momentum has a very different effect when travelling towards you than when travelling away!

Momentum is also a quantity which is *conserved*. This follows from Newton's laws of motion and, as we delve more deeply into physics, we shall come across many quantities which are conserved—they obey what are known as *conservation* laws. In the case of momentum, conservation means that, if we have a system of bodies interacting with each other but on which no external force is acting, then the total momentum of the system remains constant. By *total* momentum is meant the sum of the momenta of the different bodies *taking into account their directions of motion*. For example, consider a billiard ball with a certain momentum striking another ball at rest in a head-on collision. If we neglect the friction between the balls and the surface (i.e. assume no external force) then, after the collision, the sum of their momenta in the direction of the line of impact will be equal to the momentum of the initial ball. Alternatively, consider the firing of a gun. Initially it is at rest and there is zero momentum. After firing it, the forward momentum of the bullet must be compensated for exactly by a backwards momentum of the gun. Hence the recoil of the gun. However, because the gun has a much larger mass than the bullet, a backward momentum equal to that of the bullet is achieved with a much lower speed of recoil than that of the bullet. The same argument applies to the propulsion of a rocket in space—the backwards momentum of the ejected burning fuel is compensated for by the forward momentum of the rocket. Similarly conservation of momentum leads to the mishap that may occur when you jump from an unmoored boat onto dry land; the boat moves away from the land as you jump towards it!

There is another type of momentum which is extremely important in many aspects of physics, not least the quantum understanding of atoms and nuclei (see Chapters 5 and 8), namely *angular momentum*. It is a measure of the vigour with which a body rotates. Take the simple example of a heavy weight being rotated by hand in a circle on the end of a piece of string (see figure 2.2(a)). It is common experience that the strength needed to keep the weight rotating increases when a heavier weight is used, the circle is larger or the weight's speed is faster. The natural propensity is for the weight to shoot off in a staight line and it is the central inward force due to the hand and string—known as the *centripetal* force—which holds it in its 'orbit'. The heavier the weight, the faster it moves or the further it is away from the centre of rotation the greater the force needed and the greater the angular momentum of the weight. In the case of such a rotating weight, the magnitude of its angular momentum is simply defined as the magnitude of its momentum (mv) multiplied by its distance (r) from the centre of rotation, namely mvr. It can be seen that if any of m, v or r are increased, then the angular momentum increases. Angular momentum also applies to spinning bodies such as a top and is simply the sum of the angular momenta of all its component particles about its axis of rotation.

To rotate an object, for example a top, requires a twisting force (technically referred to as a *torque*). Just as a force changes the

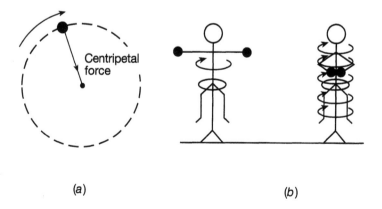

(a)   (b)

Figure 2.2: *(a) A rotating weight. (b) A rotating person.*

momentum of an object, a torque changes its angular momentum. Similarly, just as momentum is conserved when no force is acting on a system, so angular momentum is conserved when there is no torque. This is exemplified most dramatically when a person rotates on a stool which can revolve (see figure 2.2(b)). Imagine that the person is set into a spin with arms outstretched and holding a heavy weight in each hand. Now imagine the arms brought into the body. The angular momentum would reduce since the weights are now nearer the axis of rotation. But this cannot happen since there is no torque on the body—it is rotating freely—and so the only way of conserving the angular momentum of the body is for the speed of rotation to increase. This same phenomenon is seen when ice skaters start a spin with arms outstretched and then bring them into their sides.

The examples chosen to illustrate the role of angular momentum have all been simple in the sense that the rotations considered have all been essentially circular—the objects considered have rotated at a fixed distance from the axis of rotation. In many situations in nature more complicated rotational motion occurs. For example, the motion of a planet about the sun, which is held in orbit by the gravitational force between them, is elliptical, not circular. Suffice it to say here that, although more complicated in detail, the nature of such motion can again be readily understood in terms of angular momentum and its conservation.

2.4 Work and Energy

Work is a very familiar concept! At a personal level it is the energy expended in carrying out a physical task. For example, work is done when a supermarket trolley is pushed along. Technically this *work* is defined simply as the product of the pushing force and the distance over which the trolley is pushed. (The standard unit of work and energy used in physics, called the joule (see section 3.3), is the product of the unit of force (1 N) multiplied by the unit of distance (1 m).) Such a definition coincides readily with our perception of doing work—the harder we push and/or the greater the distance covered, the greater the work done. Let us explore this example a little further. Consider first the very simple and ideal situation where there are no frictional forces acting and that the trolley is on a large plane surface. Starting from rest and exerting a constant force on the trolley means that it will move faster and faster—it continually accelerates with an acceleration given by Newton's Second Law (see section 2.2). Suppose the trolley is pushed over a certain distance and then left to its own devices—no friction and no pushing force. Since there are no forces acting on it, other than gravity holding it onto the ground, it will move along steadily in a straight line with whatever speed it reached whilst being accelerated by the pusher.

The trolley clearly possesses energy by virtue of its motion. This is called *kinetic energy* and is exactly equal to the work that has been done by the pusher. It is proportional to the mass of the trolley and the square of its speed. More precisely, denoting the kinetic energy by E, for an object of mass m and speed v it is given by

$$E = \frac{1}{2}mv^2 = \frac{p^2}{2m}$$

where p ($= mv$) is the magnitude of the body's momentum. Intuitively one would expect kinetic energy to depend on these two quantities; the faster something moves and/or the more massive it is the more energy it will have. We have here a very simple example of another important conservation law—the *law of conservation of energy*. The energy expended by the pusher in doing work is conserved as kinetic energy of the moving trolley.

Consider now the more realistic situation when the trolley experiences friction as it is pushed along. As we saw in section 2.1, to push the trolley along at a constant speed requires a steady force which balances the retarding force due to friction. Here the pusher is continually doing work and yet the kinetic energy of the trolley is not increasing. So, if energy is conserved, where has the energy input from the pusher gone? The answer is that in overcoming the frictional force heat is generated—rub your hands together and you will at once feel such heat. Heat is another form of energy, as is evident from the working of steam engines, and the energy in the heat generated by friction is exactly equal to the energy expended by the pusher. Heat energy, as will be discussed in Chapter 3, is a form of *internal* kinetic energy associated with the motion of the atoms and molecules in a substance. Energy can take many forms. For example, imagine a car travelling along and suddenly having the brakes applied. Again there are frictional forces at work and, on application of the brakes, there is heating of the tyres and the road surface. There is also a screeching noise and the emitted sound carries away energy in the form of oscillations in the atmosphere. The effect of friction might be so large that sparks are emitted and then some energy is in the form of light. In this example the total energy in the heat, sound and light developed is equal to the kinetic energy of the car just before the brakes were applied.

There is also another perspective on energy which it is important to understand. Suppose you lift a weight a certain distance upwards. The weight is stationary before and after it has been lifted and so at the end of the operation it has no kinetic energy. Yet work has been done and energy expended in lifting it. Where has that energy gone? The answer is that it has been stored. If the weight were allowed to fall back to its original position it would clearly then have kinetic energy and that kinetic energy would be found to be exactly equal to the work done against the force of gravity in lifting it from this position. In its raised position the weight has the potential to release energy (if it is allowed to fall) and this stored energy is referred to as *potential energy*. Potential energy is contained in a compressed spring, an extended elastic band, the fuel in a petrol tank, food, explosives and so on. So in considering the conservation of energy both kinetic energy (related to motion of some kind) and potential energy (stored energy) must both be taken into account. It should be stressed here that it is only meaningful to talk about potential and kinetic energy in a relative sense. In the foregoing example, the weight in its original position still has potential energy since it could be allowed to fall to an even lower position. The important quantity is the *change* in potential energy in moving between the original and the raised position. Similarly, with the supermarket trolley, the speed involved in specifying its kinetic energy is measured relative to the ground and if the study were conducted on a steadily moving walkway would be measured relative to the walkway.

The law of conservation of energy in the form in which it has been described holds with great accuracy for all everyday phenomena. However, the reader should be warned that when relativistic effects are taken into account, in particular the fact that mass itself can be converted into energy—witness the atomic bomb—then this conservation law has to be modified to take this convertibility into account and we finish up with the *law of conservation of mass-energy*. This will be discussed in section 7.5.

2.5 Oscillating Systems

One form of motion pervades physical processes—oscillatory motion. We have just mentioned sound and light. Sound, as is well known, is produced by oscillating, or vibrating, systems—guitar strings, drum skins, air oscillating in a flute or trumpet, the vocal cords which produce our voices and so on. Similarly, as will be discussed in detail in Chapter 4, light is simply the manifestation of oscillating electric and magnetic fields (in conjunction, referred to as *electromagnetic* fields). For that matter so are radio waves and x-rays.

Of course oscillations vary not only in the nature of the oscillating systems but also in their *frequency* and their *amplitude*. Frequency is the number of complete oscillations that take place in a unit of time; it is defined as the number of oscillations per second. It is different frequencies that account for the different notes obtained from musical instruments. Similarly the different frequencies of oscillating electromagnetic fields correspond to different colours of light, radio waves and x-rays. Amplitude is a measure of how large an oscillation is and, for example, with sound or light determines how loud or how bright they are respectively.

In considering the behaviour of solid matter, oscillations of the component atoms will again be found to be a key feature in achieving understanding. It is therefore important to explore just a little further the nature of oscillations and the simplest system to consider is what is referred to as the *simple pendulum*. Such a pendulum is illustrated in figure 2.3. It consists essentially of a small weight (bob) suspended by a thread oscillating between two

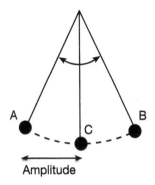

Figure 2.3: *The simple pendulum.*

points A and B. The amplitude of the oscillation is simply the maximum distance the bob moves from the centre (C) of its motion and the frequency is the number of complete oscillations it makes in a second where, by a complete oscillation is meant, for example, motion starting from A and returning to A.

The nature of the motion of the pendulum is easily understood in terms of the energy involved. At rest, the bob of the pendulum is at the point C. If it is now set into motion by a sideways push it moves, say, to A where it instantaneously comes to rest. It then accelerates under the force of gravity towards C where it has maximum speed and then slows down as it moves towards B where it again comes instantaneously to rest. It then repeats this motion but in the opposite direction and so on, if there is no friction, *ad infinitum*. At C it has maximum kinetic energy whilst at A and B it has zero kinetic energy, being stationary, but has maximum potential energy (equal to the kinetic energy at C). At intermediate positions the bob has some kinetic and some potential energy, but the total is always the same reflecting the law of conservation of energy.

The variation of the potential energy with position is shown in figure 2.4. Point C is the natural position of rest (equilibrium) of the bob where it has its lowest potential energy and, in an evocative way, it is usual to speak of it as being at the bottom of a 'potential well'. Energy is needed to move the bob to A or B and the situation is analogous to a person confined in a valley needing

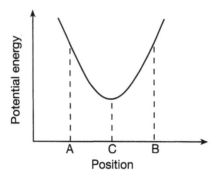

Figure 2.4: *The potential energy curve for a pendulum bob.*

energy to climb up the sides. Of course, this discussion of a pendulum refers to the ideal situation existing if there were no air resistance—it needs to be suspended in a vacuum! In reality, due to the air resistance, after being set in motion and then left to itself, the amplitude will gradually decrease as the pendulum hands over energy, as it moves, to the surrounding air molecules. Eventually, when all of the energy initially given to it has been absorbed by the surrounding air, it comes to rest at C.

It is interesting to note that, provided the amplitude is not too large, the frequency of oscillation is independent of the size of the amplitude—the increased speed of travel for a larger-amplitude oscillation exactly compensates for the necessarily increased

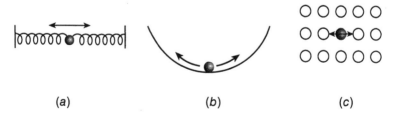

(a)   (b)   (c)

Figure 2.5: *Systems in simple harmonic motion. (a) Oscillating springs. (b) A ball bearing oscillating in a bowl. (c) An oscillating atom.*

distance of travel. Such motion is universally referred to as *simple harmonic motion* and will be frequently encountered as we plunge deeper and deeper into physics. It is exhibited (see figure 2.5), for example, by a bob held between two springs and by a ball bearing rolling in the bottom of a curved bowl through to an atom held in place by surrounding atoms in a solid (see section 3.2).

2.6 Wave Motion

Oscillating systems, if connected to the environment in some way, create waves in that environment. For example, if you oscillate your hand in a pond a wave on the surface of the water is created; an oscillating violin string creates a sound wave in the air with which it is in contact. Such waves are known as 'mechanical' waves and are very familiar. In Chapter 4, when we come to consider electromagnetic phenomena, we shall find that oscillating electric or magnetic systems can actually create waves in a vacuum; the waves are simply moving variations in the electromagnetic fields created by these systems, but no medium is needed to convey them through space. (Here it must be said, however, that until towards the end of the last century it was believed that they *were* propagated through a hypothetical medium called the aether, which was supposed to permeate the whole of space.) These latter waves will be considered later and for the moment we will confine discussion to mechanical waves.

Consider a wave created on the surface of water, for example in a pond, by an oscillating system such as a wiggling hand. It has the instantaneous form shown in figure 2.6 and that form moves forward over the pond.

The wave is characterized by three quantities:

- its *wavelength* is the distance between adjacent common points on the wave, for example, between two successive crests (as shown);
- its *amplitude* is the maximum height of the wave;
- its *speed* is the speed with which the wave moves forward.

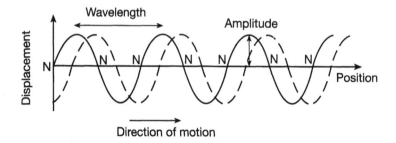

Figure 2.6: *The essentials of wave motion. (The dotted curve shows the position of the wave a little later in time.)*

Certain points on the wave (denoted by N) instantaneously have no displacement; these are known as *nodes*. Here it is important to recognize that it is the *shape* of the wave which moves forward and *not* the water within the shape; the water simply moves up and down as the wave moves along. This nature of waves is seen most clearly when a wave moves down a long string as one end is continually shaken up and down; the wave moves along the string but the string itself does *not* travel along. So a mechanical wave does not convey matter as it moves along; it does, however, by virtue of the motion of the medium in which it has been set up, convey *energy*. This is clearly seen in, for example, the damage done by large sea waves during a storm and by the energy obviously given to the ear drum when we hear a sound.

Going back to a wave on the surface of a pond, consider a fixed point on the surface. In one second a certain number of crests will move by it; this number is the *frequency* of the wave and is the same as the frequency of the oscillating system which created the wave. Clearly the speed of the wave is simply the wavelength multiplied by the frequency—the number of wavelengths that pass by in a second:

$$\text{speed} = \text{wavelength} \times \text{frequency}.$$

Such a surface wave is known as a *transverse* wave since the motion of the particles of water (up and down) is perpendicular to the direction of motion of the wave. On the other hand a sound wave is *longitudinal* in that the particles of air conveying the wave move backwards and forwards in the same direction as the motion of the wave. This follows because the vibrating string, vocal cord, loudspeaker diaphragm etc producing the sound pushes the air in its neighbourhood backwards and forwards. The situation is illustrated in figure 2.7 for a sound wave created by a loudspeaker. The shape of the wave now measures how the *density* of the air varies as the wave moves along. Peaks correspond to high density (the air molecules are squashed together) and troughs to low density (the air molecules are well separated). Here it should be stressed that in practice the sound is not confined to a beam, as illustrated, but goes out in all directions like ripples created by a wiggling hand in a pond.

Sound waves have a speed of around $340 \, m \, s^{-1}$ (760 miles per hour) in air, but travel faster in liquids and even faster in solids. Waves on the surface of a liquid travel very much more slowly whilst, at the other extreme, the speed of an electromagnetic wave (see section 4.4) is $3 \times 10^8 \, m \, s^{-1}$ (or 186,000 miles per second).

There are two important wave phenomena which it is appropriate to mention here and which will be referred to later. The first is the *Doppler effect* which occurs when the source of a wave motion is

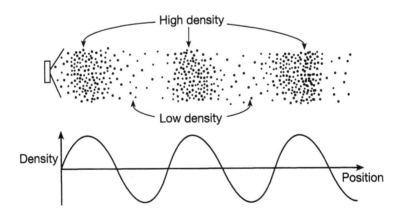

Figure 2.7: *Propagation of a sound wave in air from a loudspeaker.*

moving. For example, if the source is moving away from the observer then the wave motion is 'stretched out' and occupies more space than it would have done if the wave source had been stationary. This means that the wavelength appears to be longer and, remembering the relation between wave speed, frequency and wavelength, the frequency appears to be lower. Conversely, if the source is moving towards the observer, the wave is squashed up, the wavelength is shortened and the frequency appears to be higher. This effect is well known in the context of sound when, for example, the note emitted by a police siren drops in frequency (the note becomes lower) as the vehicle comes towards an observer, passes by and moves away.

The second phenomenon is known as wave *interference*. If, for example, the paths of two identical waves cross then the displacement caused by each of the waves separately will combine to create a single total displacement. For example, the situation may arise that at some points the trough of one coincides with the crest of another and the result is that there is zero net displacement. We then have what is known as *destructive interference*. At other points two crests or two troughs can coincide, leading to double-sized crests and troughs. This is known as *constructive interference*. Clearly the precise nature of this interference depends on the disposition of the waves. It can be readily observed in a bath by wiggling two hands in the water and seeing how the two resulting waves interfere with each other.

The waves we have been discussing are known as *travelling waves* since the wave shape travels along in the medium. However there can also be what are known as *standing waves*. These occur when the medium in which the wave is travelling is confined in some way. The simplest example is a string (e.g. a violin string) fixed at two points (figure 2.8). Since the string is fixed at each end, these endpoints of the string are stationary and must, therefore, be nodes. It then follows that the only sort of wave that can be set up on the string is one in which an integer number of half wavelengths exist between the two endpoints. A few examples of this are shown in the figure. Clearly the wave on the string does not move forwards or backwards; it is stationary or *standing* and is simply a vibration of the string. It, in fact, results from the

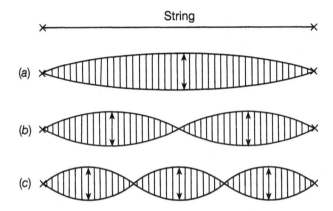

Figure 2.8: *Standing waves on a vibrating string: (a) one half wavelength; (b) one wavelength; (c) three half wavelengths.*

interference between two travelling waves moving in opposite directions on the string and being reflected backwards and forwards at the fixed ends of the string.

We have seen that the frequency of a wave is related to the wavelength and so it follows that only certain frequencies are allowed for a standing wave. The basic frequency, produced for example by a violinist, is by a standing wave of the form (a) and higher frequencies are produced by simply shortening the vibrating portion of the string by use of the fingers. He/she sometimes also produces standing waves of the form (b) by gently touching the string in the position of a node, so bringing that point to rest. Other stringed instruments (cellos, harps, pianos) similarly involve the setting up of standing waves on fixed lengths of string. On the other hand, wind instruments (trumpets, flutes, oboes) set up standing waves in fixed lengths of vibrating air columns. It should be noted here that in all instruments the 'basic' or 'fundamental' frequency is also accompanied by small contributions from higher-frequency standing waves. These are known as *harmonics* and the different mixtures of these harmonics are responsible for the widely different tones emanating from different instruments playing the same basic note. Standing waves can also be set up on the surface of a liquid confined in, for example, a tumbler. The concept of standing waves is of immense importance in physics as will become clear when we go on to consider quantum phenomena in later chapters.

2.7 Moving Forward

In the foregoing paragraphs various simple types of motion have been considered—motion in a straight line, in orbit, oscillatory and wavelike. All such motions are conditioned by the forces acting on the system being considered and on the nature of the system. Important features of the motion, which will occur frequently later on, are momentum, angular momentum and kinetic and potential energy. Of course many complicated types of motion can occur in the material world but much can be understood in terms of the simple concepts just introduced. The next step is to consider the mechanical and thermal properties exhibited by matter as a prelude to understanding these properties at a more fundamental level.

CHAPTER 3

THE NATURE AND BEHAVIOUR OF MATTER

Some Mechanical and Thermal Properties

3.1 Atoms and Molecules

The idea that everyday matter in all its forms consists of atoms derives from the work of the Greek philosophers Leucippus and his pupil, Democritus, in the fifth century BC. Atoms are the smallest entities of a pure substance—a chemical element—that can exist. The lightest, and simplest, is the hydrogen atom and one of the heaviest is the uranium atom. It is now known that atoms are not hard rigid billiard-ball-like entities (although they will be represented like that in diagrams), but have what can only be called a 'fuzzy' structure. Roughly speaking they have masses in the range from 10^{-27} kg to around 10^{-25} kg and diameters around $(1–5) \times 10^{-10}$ m. They are very, very light and very, very small!

It is clear that there must be an attractive force between atoms so that matter holds together and does not fragment into its component parts. The origin of this force depends on the detailed structure of atoms and this will be discussed in Chapter 5. Suffice it to say here that this structure is conditioned by quantum mechanics and involves electrical interactions. It is in terms of these that the nature of the force can be understood. This force is experienced only over relatively small distances—a few atomic diameters. Further, when the atoms get very close to each other, the force gradually changes from being attractive to being *very* repulsive. This means that at some point the force goes through zero (see figure 3.1(a)) so that two atoms could in principle be stationary with respect to each other at this equilibrium separation. Moving away from this point in either direction, the interatomic force acts to bring the atoms back to it—moving

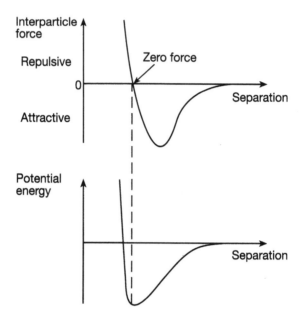

Figure 3.1: *The qualitative nature of (a) the interatomic or intermolecular force and (b) the associated potential energy.*

closer, the repulsive force pushes them apart; moving further apart, the attractive force pulls them together.

An alternative and useful way of describing this force is in terms of the potential energy experienced by the two atoms. Imagine that they are so far apart that they can only just feel the attractive force. In this position they have potential energy in that the force can pull them closer and they will gain kinetic energy. This is exactly similar to a body above the earth having potential energy because of the attraction of the earth's gravitational force (see section 2.4). As the two atoms move towards each other the potential energy eventually reaches a minimum at the equilibrium point where the attractive and repulsive forces exactly cancel one another. Pushing them closer together against the repulsive force increases the potential energy because, if released, under the influence of this force, they will move back to the equilibrium point acquiring some kinetic energy in the process. This variation of the potential energy with interatomic separation is shown in figure 3.1(b) and should be compared with the potential energy curve for simple harmonic motion shown in figure 2.4. There is, again, a 'potential well' although of a different shape. The situation is similar except that in terms of the 'valley' concept there is now a very high and very steep mountain on one side and a low hill on the other. In other words, from the equilibrium situation at the bottom of the valley more energy is needed to move the atoms closer together than to move them further apart. Indeed the particles could be completely separated if sufficient energy is provided for the low hill to be surmounted; this is the energy difference between the top of the low hill and the bottom of the valley. It is referred to as the depth of the potential well. If they are given energy, but not enough to surmount the hill, then they will oscillate about the equilibrium position alternately moving between equal heights up the low and the steep hill.

The effect of this interatomic force is that atoms tend to cluster together in small and sometimes large groups known as *molecules*. For example in hydrogen gas the atoms rarely stay alone but cling together in pairs—hydrogen molecules. This pairing happens for many other elements. On the other hand, in some elements such as helium the force of attraction is so weak that the atoms do not attach themselves to each other; elements like this (argon and krypton are other examples) are referred to as *inert*. Beyond this, different types of atom join together easily to form molecules— sodium and chlorine atoms together form a molecule of salt; two hydrogen atoms and one oxygen atom form a molecule of water. At the other extreme protein molecules have literally tens of thousands of atoms of, for example, oxygen, hydrogen, carbon and sulphur joined together into an extremely large molecular package.

Whatever the size of a molecule it will also exert an attractive force (with a repulsive core) on neighbouring molecules which has the same sort of shape and size, and for roughly the same reasons, as the force between atoms shown in figure 3.1. As has been said earlier such an interatomic or intermolecular force must in fact be expected on common sense grounds. Matter obviously holds together and there must be a force of attraction between the component atoms or molecules to ensure this. When you try to compress or extend a piece of solid matter the difficulty of doing this reflects the strengths of the repulsive and attractive features of this force in so far as the atoms or molecules are being pushed together or being pulled apart. It might be thought that the gravitational attraction between them is sufficient to hold them together. In fact, although this attraction is there, because its strength depends on the masses of the particles, which are miniscule, it is far too weak to play any part. For our purposes, in discussing the properties of different forms of matter, it is then sufficient at this stage to regard the matter as simply an assembly of atoms or molecules interacting with each other through the interatomic or intermolecular force. For brevity, in the following, we shall simply speak of molecules and the intermolecular force regarding an atom, as it were, as the simplest possible molecule.

3.2 The Particulate Nature of Gases, Liquids and Solids

Everyday matter exists in three basic forms—*solids*, *liquids* and *gases*—and it is fairly easy to conjecture the essential particulate structure of these three forms.

Solids. The fundamental difference between a solid and a liquid or a gas is that a solid is, generally speaking, a rigid structure. The implication is that each of the component molecules is held roughly in the same place—the equilibrium position—by the intermolecular force. The molecules are very close together (typically 10^{-10} m) and are arranged in ordered ways (see figure 3.2(a)) extending over large numbers of particles. There are many patterns (or lattices) in which this can happen and this is reflected in the many examples of crystalline structure. Sometimes, however, in what are called *amorphous* solids, this order extends only over a few molecules, and the solid consists of myriads of very tiny crystals—*crystallites*—arranged in a random way.

Also, difficulty in forming crystals occurs for some molecules. For example, in solids known as *polymers* the component molecules

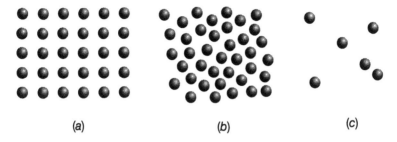

(a)   (b)   (c)

Figure 3.2: *Instantaneous views of molecules in (a) a solid, (b) a liquid and (c) a gas.*

are extremely complicated and irregularly shaped and are referred to as *macromolecules*. Packing them together in a tidy crystalline array is just not feasible. Finally, it must not be thought that the component molecules in a solid are at rest. Each molecule is always vibrating about its equilibrium position. It moves backwards and forwards between the sides of the valley of the relevant potential energy curve but does not escape from it.

Liquids. In a liquid the molecules are nearly as close to one another as in a solid but they frequently break away from their equilibrium positions. Generally speaking, each molecule is bonded to a few neighbouring molecules. However, the motion is much more complicated than in a solid and, as well as vibrating in relation to each other, the particles are also moving around in a very disordered way; an instantaneous impression of the way in which molecules are distributed is shown in figure 3.2(b). The nature of this motion is reflected in the lack of rigidity of a liquid. Its effect was dramatically demonstrated by an English botanist, Brown, in the early 19th century, who studied, microscopically, the dancing movement (Brownian motion) of grains of pollen suspended in water due to their collisions with the water molecules. It is, of course, difficult for the component molecules of a liquid to move about freely because of the drag of the attractive force and because of the very frequent intermolecular collisions—rather like people trying to move around in a crowd. This difficulty manifests itself as the phenomenon called *viscosity* which is the name for 'resistance to flow' exhibited to the extreme in substances such as treacle.

Gases. These have very low density (density is the mass per unit volume) compared with liquids and solids—they can even be 'lighter than air' (e.g. think of helium in a helium balloon). This means that the molecules are few and far between (see figure 3.2(c)); they are well separated (typically 3×10^{-9} m) compared with a solid or a liquid and, for most of the time, are too far apart to experience the intermolecular attractive force. The molecules are moving around in a random way all the time (typically with speeds around 10^2 m s^{-1}) but collide with each other from time to time. The distance they travel on average between collisions is known as the *mean free path*; in the air that surrounds us this is a few hundred times the size of a molecule. A gas can easily be compressed compared with a liquid or solid and, given the structure just envisaged, this is to be expected; there is plenty of empty space within the gas in which to bring the molecules together. It must also be recognized that the continual pounding of the moving molecules on the surface of any container exerts an outward force on it; this is simply what we know as the *pressure* due to the gas. The magnitude of the pressure is defined as the force exerted per square metre of the surface.

Why at ordinary room temperature some things are gases, others liquids and others solids relates simply to the strength of the intermolecular attractive force. There are two competing features to consider: the motion of the individual particles and the attraction of the intermolecular force. The motion tends to keep the molecules separated from each other whilst the force tends to draw them together. In gases the motion wins (the force is weak); in solids the force wins (the force is strong) and in liquids we have what might be called the 'halfway house' when both activities are equally important.

3.3 Internal Energy, Heat and Temperature

An important characteristic of all matter, as described above, is that its molecules are in continuous motion, either travelling around, as in gases and liquids, or simply vibrating, as in solids. They may also possess energy by virtue of internal vibrations or rotations. Clearly they possess kinetic energy because of these different activities. So, any substance possesses an *internal energy* due to this motion of its component molecules. They will be travelling around in gases and liquids with all sorts of speed and in all sorts of direction, and in a solid the vibrations will again be very varied in their amplitudes and directions. These motions, in other words, are entirely random in nature.

However, in all cases there will be an *average* value of this kinetic energy per molecule which is simply the total internal kinetic energy of a piece of matter divided by the number of molecules in it. This quantity is proportional to the *temperature* of the substance. Temperature, in other words, is simply a measure of the kinetic energy of the molecules of a substance: the higher the temperature the faster they are moving about; the lower the temperature the stiller they become. This latter statement has the inevitable implication that there must be a lowest possible temperature—that at which all the component molecules are at rest and have no energy. (In Chapter 6 we shall find that because of quantum effects this state of affairs is not quite attainable.) This temperature is known as *absolute zero* and, on the Celsius scale (freezing point of water, 0 °C, boiling point, 100 °C), is −273.16°C. In physics it is usual to use what are known as *absolute* temperatures, first introduced by Lord Kelvin in the latter part of the 19th century, on which the basic temperature interval (known as one kelvin or 1 K) is identical with that on the Celsius scale (i.e. 1 K = 1 °C). However, zero temperature on the Kelvin scale is taken to be the absolute zero so

absolute temperature = Celsius temperature + 273.16.

The boiling point of water, for example, is then 373.16 K since, on the Celsius scale, this boiling point (100 °C) is 373.16 degrees higher than absolute zero. It must be stressed here that temperature is a physical quantity that only has meaning at a statistical level—it is related to the average kinetic energy of a *large* number of molecules. It has no meaning to talk about the temperature of a single atom or molecule or, for that matter, of a small number.

In the light of the concept of internal energy, let us now return to the law of conservation of energy (section 2.4). There, in considering the conservation of energy in pushing a trolley or when a car brakes, it was pointed out that the heat created through friction was embodied in the increased kinetic energy of the atoms and molecules involved leading, as we have now understood, to a rise in temperature. Another way of raising the temperature of a body is for it to derive heat from a hotter body; we then think of heat flowing from the hotter body to the cooler body—energy is transferred between them. Such a transfer of heat can, in fact, take place in three quite different ways. *Conduction* (see section 3.5) is the process in which heat travels, for example, from the hot end of an object (e.g. a rod) to the cold end; there is no transfer of matter along the rod, just a transfer of energy. On the other hand with *convection*, in which, for example, hot air rises or hot liquid moves to the top in a heated pan, the substance as a whole moves. Finally, there is *radiation*, in which energy travels across space as an electromagnetic wave (see section 4.4) and heats up the objects on which it falls, for example the heat we experience from the sun or the heat generated in a microwave cooker.

If heat is supplied to a body we have seen that its internal energy increases. There is also another possible result: the body may actually do some work. Suppose gas in a cylinder confined by a piston is heated. There must be a force on the piston to withstand the pressure of the gas. When heated, because of the increased kinetic energy and, therefore, higher speeds of the molecules, the pressure on the piston will increase. The piston will be pushed out against the restraining force, thereby doing some work as in a steam engine. So the heat energy supplied can be transformed into two forms of energy—internal energy and external work. The law of conservation of energy then tells us that

> heat energy supplied = increase in internal energy
> + external work done.

This is simply a particular formulation of the law of conservation of energy but it is also dignified by being known as the *first law of thermodynamics*!

Thinking of this in practical terms means that not all of the heat energy supplied (say to a steam engine) can be converted into useful external work. Some, inevitably, increases the internal energy and, therefore, the temperature of the system. On the other hand the reverse process, in which work is converted into heat, can be carried out with 100% efficiency as was first shown by James Prescott Joule (whose name is now used as the unit of energy) in the middle of the 19th century. There is here a fundamental asymmetry: a given amount of work can all be converted into heat, but a given amount of heat cannot all be converted into work—there is always some loss of heat to the surroundings. This asymmetry leads us on to what is known as the *second law of thermodynamics*.

3.4 The Second Law of Thermodynamics

As has just been discussed, when a hot body is placed in contact with a cold body then the cold body heats up and the hot body cools down: heat flows from the hot body to the cold body. The reverse process is never experienced. Of course, with a refrigerator heat *is* extracted from the cold contents and flows into the warm room. However this is only achieved by using energy via the refrigerator motor; the process does not happen naturally—work has to be done. The foregoing is essentially a simple formulation of the second law of thermodynamics, which states that heat always flows from a high-temperature body to a low-temperature body unless work is done by an outside agent.

More insight into the nature of this law is obtained by looking at the microscopic situation. Starting with two bodies at different temperatures, the component molecules in the hotter body have higher average energies than those in the colder body; molecules with different average energies are separated. This means that there is some *order* in the system in that the molecules with *higher* average energy are separated from molecules with *lower* average energy. It is similar to the order achieved when you have one box containing red buttons (say) and another containing blue buttons. When the two bodies are placed in contact with each other the flow of heat means that eventually they reach a common temperature and all the molecules have the same average energy; there is less order (more disorder) in the system. With our button analogy, it is equivalent to mixing all the buttons in one box. In other words, the second law of thermodynamics can be taken to state that the flow of heat in a system is always in such a direction as to increase the disorder of the system. A physical quantity that expresses the amount of disorder in a system is what is called its *entropy*. Entropy is simply related to the probability of a given state of the system existing—the lower the entropy the less probable is the state. For example, consider a single box containing both blue and red buttons. If it is shaken, it is possible, but *very improbable*, that you will end up with an ordered (low-entropy) state in which the blue buttons are on one side and the red buttons on the other. It is, in contrast, *highly probable* that the buttons will become very mixed up and highly disordered (high entropy). Therefore, another formulation of the second law is that in all physical processes taking place in a closed system, i.e. with no external influences, the entropy (disorder) always increases. This law is a statistical law and so it must be recognized that it *is* possible, although negligibly so, that a system *might* end up in a more ordered state with a resultant decrease in entropy. The likelihood of shaking a button box and finding the blue and red buttons completely separated is extremly small; how much more so for interacting pieces of matter containing 10^{24} or more particles!

The whole universe is subject to the second law, which means that its entropy and therefore its state of disorder is continually increasing. Of course, order is all around us in the form of furniture, machines, roads, buildings, indeed ourselves etc, but this order is only achieved by vast expenditure of energy which leads to ever increasing disorder elsewhere. Eventually (but an inconceivably long time ahead) it might be thought that there will be complete disorder everywhere and a common temperature throughout the universe; the so-called 'heat death' will be upon us! That being said, there are residual uncertainties about this conclusion when it is put into the context of the expanding universe (see section 10.3) and the possible eventual contraction of the universe leading to the 'big crunch'.

This continual increase of entropy with time in the universe has another implication, namely that it indicates the direction in which time is flowing—the 'arrow of time'. Make a video of a china plate (a very orderly state) being dropped on the floor and breaking up into smithereens (a very disordered state)—a process in which the disorder, and therefore the entropy, clearly increases. Then run the video backwards. Clearly the reassembly of the plate shown in this backward run does not accord with our everyday life; time just does not run in that direction. Our experience of time runs in the direction in which entropy increases. However, this discussion is in the context of the behaviour of very complex entities. At a fundamental level involving elementary particles the direction of time is obscure and, in section 9.4, we shall come back again to the nature of physical laws when time is 'reversed'.

However, let us for the present stay with everyday matter and its behaviour, which is governed by, among other things, the two laws of thermodynamics just discussed. In the following some of the important properties of solids, liquids and gases will be briefly described and explained.

3.5 Solids and their Behaviour

As has been said, solids are essentially rigid structures in which the particles vibrate about localized positions in an ordered array (often some form of crystal lattice). This perspective on the structure of a solid enables a number, but not all, of the different characteristics of solid matter to be understood.

Rigidity and Elasticity. Although solids are essentially rigid, most can be stretched, squashed and twisted. In such processes the component molecules are displaced from their equilibrium positions. Here, incidentally, it should be noted that, because of the shape of the intermolecular potential energy curve (figure 3.1), it is much harder to squash a solid than to stretch it. In terms of the valley analogy it is harder to make lateral progress climbing the steep mountain than it is climbing the low hill. If the forces producing such distortions are removed then, provided the distortion is not too great, the solid will revert back to its original shape; the interparticle forces bring the particles 'back into line'. The extent to which a solid is distorted by a given external force is, as might be expected, generally proportional to the strength of the force, the constant of proportionality being known as an *elastic modulus*. For solids, three such moduli are measured. They give information about the extent to which a solid can be stretched, twisted or squashed. Their values can also be used to estimate the strength of the relevant interparticle forces. Elastic properties of solids vary widely and an extreme situation is that of rubber which consists of molecules in the form of very long chains of carbon and hydrogen atoms. In equilibrium these chains are in a 'higgledy-piggledy' curled-up state and the striking elastic properties of rubber result from the way in which these long-chain molecules can be elongated when the rubber is stretched and then revert to the original shape when released.

If a solid is distorted too much then it will break. However there is much variation in how this comes about. Some metals (known as *ductile*), for example, can be distorted very much, and will retain that distortion, before finally breaking. Other metals (e.g. cast iron or glass) are *brittle* and after small distortion suddenly break. Then, again, plastic materials and rubber, which consist of long-chain molecules that can slide over each other, can be stretched very considerably before breaking.

Thermal Expansion. When a solid is heated it expands and this can again be understood in terms of the shape of the potential energy curve (figure 3.1). It has been mentioned that particles in a solid are vibrating about their equilibrium position. The higher the temperature the more energy this motion has and so the higher the system can climb up the potential energy curve (see figure 3.3). For a given energy the extremes of the motion are limited by the shape of the curve as shown in the figure. Because of the lopsided nature of the curve (steep mountain and low hill) it is clear that the higher the energy the greater is the average separation (centre point of the vibration). Hence, expansion takes place.

Melting. The phenomenon of melting (or fusion) is not well understood in detail but can be associated with the fact that as the temperature of a solid is increased so the amplitude of the particle vibrations increases. This means that a 'loosening-up' of the solid is taking place and it is at least qualitatively understandable that it makes a transition to the liquid state—what is called a *phase change*. When a solid reaches the melting temperature it remains at that temperature whilst the melting takes place and there is no

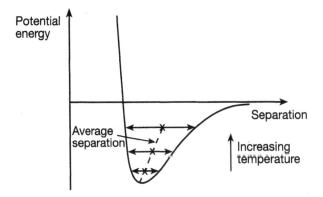

Figure 3.3: *Vibration amplitude for different energies.*

rise in temperature until the melting is completed. The heat provided to melt a solid is known as the *latent heat of fusion*. It has a much smaller value than the *latent heat of sublimation*, which is the heat needed to change a solid directly into a vapour. This is simply because, whilst in the liquid state the particles are in and out of the potential well, in the gaseous state they are completely separated (figure 3.2(c)) and all have had to receive sufficient energy to surmount the depth of the potential well (the low hill).

Thermal Conductivity. This is the process in which heat energy is conducted through a substance and can be understood qualitatively as follows. Suppose a small region of a solid is heated. The temperature will rise and the molecules in that region will vibrate with increasing violence and amplitude. Because the neighbouring molecules feel the force of attraction of these agitated molecules they will be dragged into more violent motion themselves and so on through the solid. The disturbance will therefore be transmitted through the solid, molecules further and further away from the point of heat input will vibrate more energetically and the temperature there will therefore rise. In its travels the disturbance will meet obstacles, for example the surface of the solid, or imperfections and impurities, and taking the effect of these into account determines how good a conductor of heat a solid is. However, in dealing with metals, which are usually much better conductors of heat than non-metals, there are clearly other factors in play to account for this. These relate to the actual structure of the component atoms and to the fact that metals are also good conductors of electricity. This aspect of thermal conductivity is referred to in section 6.2.

3.6 Liquids and their Behaviour

Liquids, as was indicated in section 3.2, are intermediate in structure between solids and gases. They have properties of both: for example, densities akin to those of solids but the same lack of rigidity as gases. The constituent molecules are not tightly bound together as in solids, neither are they virtually completely free as in gases. For these and other reasons, although their general behaviour can be understood qualitatively in terms of their structure, it has proved difficult to obtain very precise understanding; liquids are a very complicated form of matter. Sometimes their properties can be understood by regarding them as a modified solid, sometimes as a modified gas and sometimes as a liquid *sui generis*. In the following we just consider two important properties in a little detail.

Vaporization. If a liquid is in a partially filled container from which all the air has been evacuated, some of the molecules of liquid will occupy the space above the liquid in the form of vapour which will, of course, exert a pressure (*vapour pressure*) on the surface of the container. *Evaporation* is said to have taken place and can easily be understood. The liquid particles are moving around in the liquid with all sorts of kinetic energies, the average energy being dependent on the temperature. Some will be moving so fast that they can overcome the restraining intermolecular force and escape into the space above. The most energetic molecules escape, so leaving lower-energy molecules behind and a resultant lower temperature. (This effect is enhanced in our everyday experience when the escaping molecules from our bodies are removed by a wind leading to the 'wind-chill' factor frequently referred to in weather forecasts.) Returning to the liquid, some of the escaping molecules will also be 'recaptured' and, eventually, an equilibrium situation is established when as many escape as are being recaptured. The higher the temperature the higher the kinetic energy and so the more escape until, at a sufficiently high temperature *(boiling point)*, all the liquid can be converted into vapour. The amount of heat that has to be supplied to achieve this is known as the *latent heat of vaporization* and is very like the latent heat of sublimation referred to in section 3.4.

Surface Energy. A molecule in the bulk of a liquid is subject, on average, to no net force because it is surrounded on all sides by other molecules (see figure 3.4). However a molecule in the surface of a liquid experiences a net inward force due to the attraction of the particles immediately below the surface (see figure 3.4). This means that to increase the surface area of a liquid requires the expenditure of energy in order that molecules moving to the surface can overcome this net inward force. Conversely, the equilibrium situation—the situation of lowest energy—for a liquid is one in which the liquid reduces its surface area to a minimum. This propensity of a liquid then accounts for the phenomenon of *surface tension*, the resistance experienced when trying to increase the surface area of a liquid; for example, when blowing a bubble. This effect is also the reason why a tiny amout of liquid tends to form itself into a spherical drop, the shape with minimum surface area, rather than spreading out into a thin layer with a much larger surface area. It should be noted that both surface energy and latent heat of vaporization are related to the strength of the intermolecular force and, for this reason, these two quantities are very roughly proportional to each other.

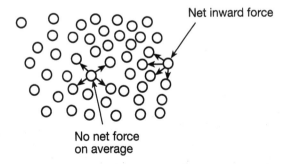

Figure 3.4: *Forces on molecules in the bulk and in the surface of a liquid.*

3.7 Gases and their Behaviour

Gases are the simplest of the different states of matter to understand. This is because the motion of the component molecules is essentially free apart from occasional collisions with other molecules.

Pressure. We have spoken of the pressure exerted by a gas on the sides of its container as being due to the continual pounding of these sides by the gas molecules. The magnitude of the pressure can be understood simply in terms of change of momentum. A molecule heading straight for the side of the container will have a certain momentum depending on its speed. When it strikes the side it will bounce back and its momentum will be reversed. This change in momentum implies that a force is exerted on the particle by the side and, in turn, because of Newton's Third Law of Motion (see section 2.2), an equal but opposite force is exerted on the side. This is the origin of gas pressure, which is simply defined as the total force exerted on unit area of the side by the gas molecules. Of course not all molecular collisions with the side are 'head on' and this has to be taken into account when working out exactly what the pressure is. It turns out that the pressure is roughly proportional to the average kinetic energy of the particles and, therefore, to the temperature of the gas. This is common experience; heat up gas in a container and eventually it will explode! It is also evident that the pressure must depend on how frequently the container sides are struck. It is possible, therefore, to increase the pressure in a gas simply by decreasing the volume of its container. For example, halve the volume occupied by a gas and its density will be doubled so twice as many molecules will be bombarding any side of the container. This means that the pressure has doubled. The same effect is experienced when a car tyre is inflated except that here the container size remains constant and more molecules of gas (air, in this case) are pumped into it, so increasing the density and, therefore, the pressure.

Condensation. Another feature of a gas is that if it is cooled then it condenses into a liquid; witness condensation of steam onto a cold window. This is understandable in qualitative terms. Cooling the gas reduces the speed of its component molecules. This means that when two molecules collide, if they are moving sufficiently slowly, the attractive intermolecular force is strong enough to hold them together. More and more collisions of this kind mean that a large number of molecules will be clinging together so the gas makes a transition into the liquid state. It is a transition from figure 3.2(c) to figure 3.2(b). In terms of potential energy, it means that molecules sink towards the bottom of the valley illustrated in figure 3.1(b). If they are confined to the valley then, for energy to be conserved, their potential energy when they were approaching each other must have been handed over to the surroundings. In other words, heat is released. This heat must, in turn, be *provided* when a liquid is turned into a vapour and is, of course, the latent heat of vaporization discussed in section 3.6. The temperature at which a gas condenses into a liquid is obviously dependent on the strength of the intermolecular force—the weaker this force the lower the temperature. Helium, as has been mentioned earlier (section 3.1), has a very weak interatomic force and temperatures only a few degrees above absolute zero are needed before it can be liquefied (see section 6.6). It should also be stressed that temperatures at which condensation takes place are dependent on the gas pressure. For example, the higher the pressure the closer are the molecules and so the easier it is for them to hang together in liquid form.

3.8 Moving Forward

The foregoing discussion has outlined how some of the main thermal and mechanical properties of solids, liquids and gases can be understood in terms of the motion of, and the forces between, their component atoms and molecules. The basic ideas are fairly simple, but it must be recognized that to work out this behaviour in fine detail can be extremely complicated. Of course, electromagnetic phenomena have hardly been mentioned at all. The reason for this is that, to understand them at the fundamental level, progress can only be made if the internal structure of atoms is taken into account. This will be taken up shortly, but first we must look at the nature of electromagnetism itself. This is the concern of the next chapter.

CHAPTER 4

EVERYDAY EXPERIENCE OF ELECTROMAGNETISM

The Relationship between Electric and Magnetic Phenomena

4.1 Electric and Magnetic Forces

Everyone is familiar with magnets and magnetism. A bar magnet has a north pole at one end and a south pole at the other—so-called because when the magnet is suspended from its centre, these poles always point toward the north and south poles of the earth respectively. This is due to the fact that the earth itself behaves like a very large magnet and exerts a *magnetic force* on the suspended magnet, always twisting it into the same direction. This, of course, is the mechanism of a magnetic compass. The magnetic force and its nature can easily be investigated by bringing two bar magnets close to each other. If one is suspended (see figure 4.1) it will be found that, when brought close to it, the south (S) pole of the other will attract (pull towards it) the north (N) pole of the suspended magnet whilst its north pole will repel it (i.e. push it away). It will also be found that the effect becomes smaller the further apart are the two magnets. The magnetic force can thus be attractive or repulsive and dies away with distance. Here it must be noted that

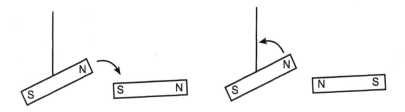

Figure 4.1: *(a) Attraction and (b) repulsion between magnets.*

as far as magnetism is concerned the *geographical* north pole of the earth is actually near to the earth's *magnetic* south pole; that is why it attracts the north pole of a magnet towards it.

Also familiar to many is the *electric force*. As remarked in section 2.1 this manifests itself most simply when, for example, an inflated balloon is rubbed against a piece of material and then held close to small pieces of tissue paper which will attach themselves to the balloon. This attachment is due to an attractive electric force between the balloon and the paper. We speak of the balloon as being electrically *charged*. There are two types of electric charge which are referred to as *positive* (+) and *negative* (−) and the unit of charge is called the *coulomb* after an 18th-century French physicist. Normally the net electric charge of an article is zero; the amount of positive charge it contains is exactly balanced by the negative charge. Here, anticipating discussion in the next chapter, it should be made clear that each atom in a substance contains equal amounts of negative and positive charge; the negative charge is carried by very light particles called *electrons* which surround the massive and positively charged *atomic nucleus*. When the balloon is rubbed against material some electrons are transferred from one to the other so that the balloon and the material each have a net charge—one because it has lost electrons (and so its net charge is positive) and the other because it has gained electrons (and so its net charge is negative). When the balloon is held near to the tissue paper it then attracts the oppositely charged particles in the tissue paper towards it and repels the like charges and the net effect of these two (unequal) forces is the observed attraction between the balloon and the tissue paper.

Analogously with magnetic poles and their associated forces, like charges repel each other and unlike charges attract. This force of repulsion or attraction depends on the sizes of the electric charges involved and it dies away as the distance between two interacting charges increases. This dependence on distance obeys an *inverse square law*. That is to say if the distance between the charges doubles, the force decreases by a factor of two squared, namely four; if it trebles, the force decreases by a factor of nine and so on. Here the reader is reminded that, as remarked in section 2.1, the fields of influence surrounding electric charges and magnets are referred to respectively as *electric* and *magnetic fields*. Fields are defined by their *strength* and their *direction* (i.e. they are *vector* quantities) at every point. For example, with an electric field the strength and direction are those of the force experienced by a positive unit electric charge (one coulomb) placed at a point.

Another important feature of electric charge is that it is *conserved*. This means that in an isolated system the electric charge of the system remains constant: it can neither be created nor destroyed. We shall see later on that charged elementary particles can be created or destroyed but *always* in association with another particle having an exactly equal but opposite charge so that the net charge created is zero.

It must be stressed that the above analogy between the electric and magnetic force is highly superficial. Whilst electric charges can exist in isolation, north and south poles do not; they always occur together, as what is called a *dipole*, for example at each end of a magnet. However, in passing, it must be said that some possible fundamental theories (see section 9.5) do suggest that isolated poles (referred to as *monopoles*) might exist. However, they would be immensely heavy and, so far, none have been detected. Certainly for our purposes at this juncture, and as we shall see in section 4.3, magnetism should be regarded as a phenomenon which is not intrinsic but derives from the actual motion of electric charges.

4.2 Electric Potential and Electric Current

Consider a unit (i.e. 1 C) negative charge in the vicinity of a fixed positive charge as in figure 4.2. An attractive force is clearly experienced by each charge and in moving the negative charge from A to B work has to be done against the attractive force. This means that the potential energy (remember the discussion in section 2.4) of the negative charge has increased. The difference in the potential energy between A and B is proportional to what is known as the *potential difference* between the two points which is measured in *volts*. The potential difference is simply the work needing to be done by an outside agency to move a unit charge

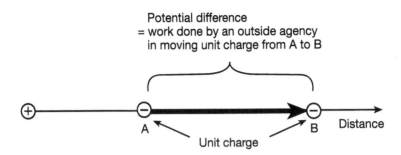

Figure 4.2: *The potential difference between two points A and B.*

from A to B. The volt is a unit which is familiar in everyday life from, for example, torch batteries which are usually rated at 1.5 V. Of course, if the negative charge is placed at B then, if it is free, it will move towards the positive charge under the influence of the attractive force and its potential energy will have decreased. So, to conserve energy, it will either have gained kinetic energy or have done some external work.

Here it is convenient to introduce an important unit of energy which is used extensively in atomic and nuclear physics. It is known as the *electron volt* (denoted by eV) and is simply the amount of work which has to be done in moving one electronic charge (magnitude $e = 1.602 \times 10^{-19}$ C) through a potential difference of one volt. Its value is

$$1\,\text{eV} = 1.602 \times 10^{-19}\,\text{J}$$

where a joule (J) is the basic unit of work and energy (see section 2.4).

A battery, having a potential difference between its terminals, clearly has the ability to convey electric charge from one terminal to the other and this can be achieved if there is a suitable conduit for the moving charge. As is well known, such conduits are electric wires and connectors of one form or another (see, for example, figure 4.3).

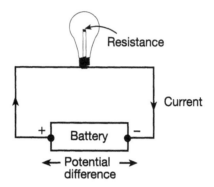

Figure 4.3: *A circuit in which current flows through a resistive light bulb.*

Wires and connectors are usually made of metal and are referred to as *electrical conductors*. It is a property of conductors that electric charges (carried by the negatively charged electrons already mentioned) can flow through them with more or less difficulty with speeds of around 10^6 metres per second. This flow of electrons is referred to as an electric current and although the electrons obviously flow towards the positive terminal of a battery, since they are negatively charged, it is historical convention to speak of the current as flowing in the opposite direction!

The inhibition of electron flow is due to collisions of the electrons with imperfections and impurities in the conductor. It is referred to as *electrical resistance* and one speaks of *good conductors* (e.g. copper) and *bad conductors* (e.g carbon) according to the size of the inhibition. For a given potential difference the size of the current (which is measured in *amperes* or, for short, *amps*—the number of coulombs flowing per second) is, as might be expected, proportional to the size of this difference. It is also inversely proportional to the size of the resistance (measured in *ohms*). Thus, we can write (Ohm's Law)

$$\text{current} = \frac{\text{potential difference}}{\text{resistance}}.$$

Of course, there are some substances, known as *insulators* (e.g. rubber, plastics), which do not allow electrons to flow through them at all unless the potential difference is *very* large indeed. In this case there is usually a violent discharge of electricity, the extreme example of this being a flash of lightning, and the insulator breaks down.

As the charged electrons move through resistive material under the influence of a potential difference they experience a sort of friction and, as might be expected, this leads to the generation of heat and, sometimes, light. This is our everday experience with current flowing through the filament of an electric light bulb or the red-hot wire of an electric fire. The amount of heat and light generated in unit time which, by the law of conservation of energy, is the amount of energy consumed in that time, is called *power*. It is the work done in unit time in conveying charge across a potential difference and, remembering our discussion above, is simply the product of the potential difference (measured in volts) and the charge transported across the potential difference in unit time (i.e. the current flow—measured in amps). The value of this product is measured in *watts*, a unit which is familiar in specifying the power rating of domestic appliances. It may happen that the amount of heat generated by current flowing through a conductor is enough to melt it! This is precisely what happens when a fuse 'blows' and it is standard practice to rate fuses by the amount of current they can carry (e.g. 3, 5, 13 A) before melting and so breaking a circuit.

4.3 Magnetism and Electromagnetic Induction

Around 1820, a Danish physicist (Oersted) discovered that an electric current flowing down a wire changed the orientation of a nearby compass needle. Clearly a magnetic field was being created by the current, i.e. by the moving electrons. This was a vastly important discovery since it showed that what, at that time, had been thought to be quite independent physical manifestations— electricity and magnetism—were, in fact intimately related. It then transpired that if the electric current were flowing round in a loop of wire, the magnetic field associated with the loop was essentially the same as that due to a bar magnet, the loop behaving as though it had north and south poles. The relationship between the poles and the direction of current flow is shown in figure 4.4(a).

It also follows that, since an electric current behaves like a magnet, a current-carrying wire will experience a force when placed in a magnetic field due, for example, to another magnet. Even more interestingly, a moving electric charge, being a current, will also experience a force when moving through a magnetic field. Here the fact that it is moving must be stressed since, if it were stationary, there would be no current and no resultant magnetism. The direction of this force and its effect on a positive charge moving in the magnetic field created by placing the north and south poles of two magnets near to each other is shown in figure 4.4(b). Its effect is to force the charge to move upwards as it travels along. This is an interesting and new type of force since it only operates when the charge is in motion and the direction of the force relates to the direction of motion of the charge. It is the force that underlies the operation of, for example, electric motors, where current flows through a coil of wire free to rotate (the armature) in a magnetic field, the latter being provided either by permanent magnets or by current flowing through fixed coils.

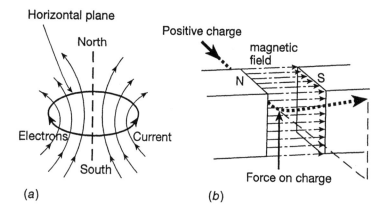

Figure 4.4: *(a) The magnetic field created by a loop of electric current; (b) the force on an electric charge moving in a magnetic field.*

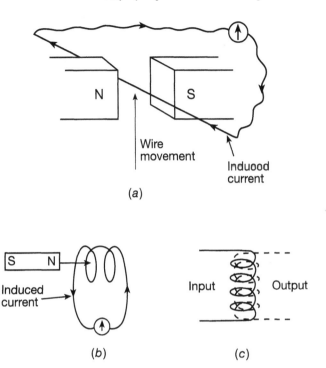

N

S

Wire
movement

Induood
current

(a)

S   N

Induced
current

(b)

Input   Output

(c)

Figure 4.5: *(a) Current induced in a moving wire; (b) current induced in a coil by a moving magnet; (c) the physical basis of a transformer.*

The considerations so far lead on to another phenomenon known as *electromagnetic induction*. We have seen that a charge moving in a magnetic field experiences a force. It therefore follows that if a wire, and therefore its component electrons, is moved in a magnetic field (see figure 4.5(a)), the electrons will experience a force, in the direction of the wire. Those which are free will therefore move along the wire under the influence of this force, so setting up an electric current, which can be measured by an appropriate meter. This was first demonstrated by Michael Faraday in 1831. Clearly the generation of an electric current from mechanical motion has had a profound effect on the development of the modern world. All the electricity which arrives in our homes, industry, transport etc derives from this simple process. It should be stressed here that the wire does not need to be moving as in the example quoted; instead the field can be moved—it is the *relative* motion of the wire and the field which is important. For example, moving a magnet into a coil of wire as in figure 4.5(b) induces a current in the wire. As can easily be conjectured, if the magnet is pushed in and out of the coil the created current will surge backwards and forwards. Such a current is known as an *alternating current* (AC) and is familiar as the type of current which comes into our homes from the generators in our electrical supply systems. These generators are, of course, massive electrical engineering structures, but the basic principle of their working is as just described. Different electromechanical structures are used in generators producing a *direct current* (DC), like the current produced by a battery, but the underlying principle is the same.

One great advantage of using AC is that the voltage can be changed at will by using a transformer. This is simply a device in which the current is passed through a coil of wire linked with another coil as in figure 4.5(c). The alternating current in the 'input' coil creates a continually changing magnetic field which induces an alternating current in the 'output' coil. The ratio of the output to the input voltage is determined by the number of turns of wire in each coil and the efficiency of the transformer is much enhanced by winding the coils round ferromagnetic material so as to concentrate the magnetic fields. Transformers have many uses. In particular, power (proportional to current multiplied by voltage) can be transmitted at a very high voltage and correspondingly low current over large distances (witness the electric pylons with their high-voltage warnings). The benefit of this is that because of the low current the wastage of energy due to heating of the transmission cables is very small. Transformers near the consumer are then used to reduce the voltage to an appropriate level. For example, in the UK power is initially transmitted at 400,000 V and stepped down for domestic use to 230 V.

4.4 Electromagnetic Radiation

In the preceding paragraphs we have seen that a steady current produces a magnetic field in its vicinity. Further, if the current is alternating—i.e. electrons in the wire are oscillating—then the associated magnetic field will also be continually changing—we have an oscillating magnetic field. However, as we have seen (e.g. figure 4.5(a)), a changing magnetic field causes a current to flow in a wire; it must, therefore, have associated with it an *electric* field which causes the electrons to move along the wire. These two fields— magnetic and electric—will, of course, be present whether or not the wire, in which the electrons are being forced to move, is there. So we have the situation that oscillating electrons produce a combination of oscillating electric and magnetic fields which propagate out into space. This phenomenon is known as *electromagnetic radiation* and, at a fundamental level, is produced by *accelerating* electric charges, i.e. charges which are continually changing their motion.

Bearing in mind the directions of the electric and magnetic forces indicated in figures 4.4 and 4.5 it is, perhaps, not surprising that the magnetic and electric fields are perpendicular to each other and also perpendicular to the direction in which they are being propagated (see figure 4.6(a)). When the alternating current is first switched on the propagation of the radiation is not instantaneous; it takes time for it to reach any point in space. It is rather analogous to the generation of a wave in a pond when a hand starts to wiggle the water at one side (see section 2.6) and a wave is propagated over the surface. The analogy is more than trivial because electromagnetic radiation is, indeed, a wave motion. However, there is a fundamental difference. With a water wave or a sound wave, the quantity that varies (water height, density of the air) is a property of the medium on or through which the wave is passing. However, with an electromagnetic wave, the varying elements are the electric and magnetic fields which are intrinsic to the wave itself and, as mentioned earlier, can be regarded as a 'stress' in space. This means that an electromagnetic wave can pass through a vacuum—it has no need of a medium in which to be propagated. The preceding discussion has been qualitative and it is gratifying that the relationship beween electricity, magnetism and electromagnetic radiation can be formulated in most elegant quantitative way; this was was first done in the 1860s by a remarkable Scottish physicist—James Clerk Maxwell.

An electromagnetic wave can thus be represented as in figure 4.6(b) (cf figures 2.6 and 2.7), where the amplitude of the wave

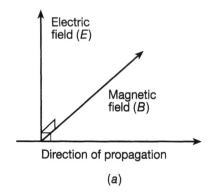

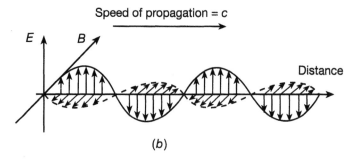

Figure 4.6: *(a) Electric and magnetic fields in electromagnetic radiation; (b) an electromagnetic wave.*

measures the strength of the electric or magnetic field at a point in space. Since the electric and magnetic fields are perpendicular to the direction of travel, electromagnetic waves (like water waves) are said to be transverse. They move through empty space at a speed (always denoted by c) given by $c = 2.998 \times 10^8 \, \mathrm{m \, s^{-1}}$ or, approximately, 186,000 miles per second. Like mechanical waves they also carry energy with them whose value depends on the strengths of the electric and magnetic fields. Their nature depends on their frequency and wavelength which (see section 2.6) are simply related to each other and the speed of the wave. Examples of electromagnetic waves are visible light, ultraviolet and infrared rays, radio and television waves, x-rays etc. A rough indication of the wavelengths and frequencies of these different types of wave is given in table 4.1.

Table 4.1: *Wavelengths and frequencies of different types of electromagnetic radiation.*

Frequency (oscillations per second)	Name	Wavelength (metres)
10^{17} upwards	x-rays	10^{-9} downwards
10^{15}	ultraviolet	3×10^{-7}
7×10^{14}	blue light	4×10^{-7}
4×10^{14}	red light	7×10^{-7}
10^{14} 10^{12}	infrared	10^{-6}–10^{-4}
10^{11}–10^{10}	microwave	10^{-3}
10^{10}–10^{9}	radar	10^{-2}–10^{-1}
10^{8}	TV/FM	1
10^{7}	shortwave	10
10^{5}	longwave	10^{3}

There is clearly a vast spread in wavelengths and frequencies and the means of generation of these different waves varies. Radio, TV and radar waves are produced using complicated electronic devices. Infrared, visible, ultraviolet and x-rays are, as we shall see later, produced using atomic processes involving transitions of electrons mainly *within* atoms. Even higher frequency radiations, known as *gamma-rays,* are produced when transitions take place within atomic nuclei themselves (see section 8.6). That being said, and in spite of the great difference in the qualities of these different radiations, it must be stressed that the waves all have the same physical nature and the differences relate solely to the differences in their wavelengths and frequencies. Perhaps the most fully studied of all the electromagnetic radiations is visible light and we now go on to consider some aspects of its nature and behaviour.

4.5 The Reflection and Refraction of Light

Visible light is characterized by a small spread of frequencies (see the above table) responsible for the different 'colours of the rainbow'—violet, indigo, blue, green, yellow, orange and red in descending order of frequencies. It was by using light that the first reasonably accurate estimate of the speed of electromagnetic radiation was made by measuring the time it took a pulse of light to travel forwards and backwards (using a mirror to reflect it back) between two hill tops. This is rather like estimating the speed of sound by measuring the time it takes for an echo to return. The major difference is that whilst the latter time can be a matter of seconds, because of the high speed of light, measurements of the order of 10^{-4} s or less are needed, requiring ingenious timing methods.

Reflection of light has just been mentioned and this is an everyday experience—for example, reflection in mirrors, shop windows and so on. Mirrors have heavily silvered backs to give virtually a 100% reflection, the light being bounced off at the same angle to the perpendicular to the surface as it hits the mirror (see figure 4.7(a)), rather like a snooker ball bouncing against the edge of the table. However it is important to note that even when a material is transparent, like glass, there is still some reflection. Some of the light is transmitted through the glass and some is reflected at the surface. This is shown in figure 4.7(b) where it can be seen that the transmitted ray of light is bent towards the perpendicular to the surface. This latter process is known as *refraction* of light. It is interesting to note here that when light is reflected from a transparent material it becomes *polarized*. That means that the electric field, for example, points in a particular direction (rather than in random directions) perpendicular to the motion of the light. It is because of this that *polaroid* sunglasses can be used to restrict glare from reflected sunlight. Polaroid is a material in which the component (rather long) molecules are arranged in parallel lines to form a sort of grid through which polarized light has difficulty in passing unless the direction of polarization is parallel to the grid. Similarly, such grids can be used to polarize unpolarized light since only light with the field in the same direction as the grid can pass through it.

The refraction or bending of light just referred to underlies the function of lenses, for example a magnifying glass, and it is important to understand the orgin of this bending. Consider the more detailed diagram of a ray of light incident on a glass surface in figure 4.7(c) where the lines across the ray again represent successive peaks in the amplitude of the light wave, the distance

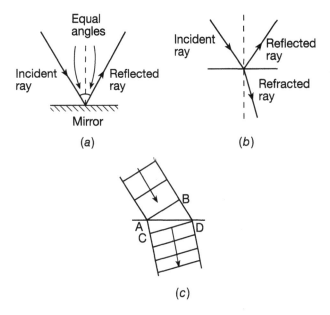

Figure 4.7: *(a) Reflection of light; (b), (c) refraction of light.*

between them being one wavelength (refer to figure 2.6). In glass, light travels at a slower speed than in air and so in the time that the end B of the wavefront reaches D the end A travels a shorter distance to C. This results in the ray bending towards the vertical as shown. The wavelength in the glass obviously becomes shorter, this resulting from the relation between frequency, wavelength and wave speed given in section 2.6 where it can be seen that the wavelength is simply proportional to the speed. The speed of light waves in glass depends on the frequency of the wave, i.e. the colour of the light, the slowest speed being for violet and the fastest for red. This, in turn, means that a violet ray is bent most and a red ray least in passing from air to glass. We speak of the light being *dispersed*.

This spreading out of different colours is seen at its most spectacular in the formation of a rainbow which results from the dispersion of sunlight through myriads of raindrops. The fact that sunlight splits up into the different 'rainbow' colours implies that it is a mixture of all these colours. Such light is referred to as *white light* since it is perceived to be essentially colourless by the eye. When white light falls, for example, onto a green object then colours other than green are absorbed by the object and only green light is reflected; the object therefore looks green. This means that a green car illuminated at night by orange-coloured fluorescent street lighting will absorb the orange light and there will be little, if any, green light to reflect back; so virtually *no* light will be reflected and the car will look black. Similarly the open sky looks blue because the air molecules tend to reflect (scatter) blue light passing through it from the sun. In turn, if we look at the sun directly when it is just above the horizon its white light will have had the blue component significantly reduced because of scattering in the air it has traversed. The residual light, having lost most of its blue component, therefore becomes reddish, leading to the magnificent sunsets often observed.

4.6 The Interference and Diffraction of Light

In section 2.6 it was mentioned that interference can take place between waves and this is a phenomenon which is well studied in the context of light waves. Consider a light wave striking an impenetrable screen with a very narrow (much smaller than the wavelength of the light) slit in it as shown in figure 4.8(a) where the vertical lines again represent peaks in the wave amplitude. Light to the right of the screen clearly originates only from the slit and this leads to *circular* wavefronts moving out as shown. The light, instead of simply moving forwards, spreads out in all directions and the process is referred to as *diffraction*. This is a phenomenon which can be similarly observed when waves on the surface of water strike a barrier with a small gap in it.

Consider now the situation which arises when there are two such slits as in figure 4.8(b). Clearly the wave fronts from the two slits will interfere with each other and the crosses indicate points at which crests of each circular wave coincide. They enhance each other and lead to what is called *constructive interference*. Similarly, where crest and trough coincide, there is cancellation and *destructive interference* takes place. If the light passing through the

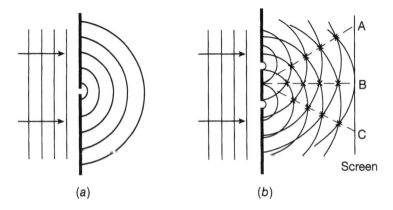

(a)   (b)

Figure 4.8: *(a) Circular wavefronts from a narrow slit;*
(b) interference from two slits.

two slits is allowed to fall onto a screen as shown then there will
be bright areas at A, B and C and dark areas in between where
destructive interference is taking place. More detailed considerations show that B is the brightest area and that the intensity of the
light in the other bright areas decreases the further they are away
from B. The pattern of light and dark areas on the screen is
known as an *interference pattern*.

Return now to the light emanating from a single slit, but consider
the situation where the slit is wider with a width of the same order
of magnitude as the wavelength of the incident light. Whereas
with a very narrow slit the circular wave is generated from
effectively one position, now there are many positions available
within the slit and so many circular waves are generated from
each point in the slit. These waves, like those from the two slits in
figure 4.8(b) will clearly interfere with each other and if the light
from this wider slit falls onto a screen a pattern of light and dark
areas is obtained; this is known as a *diffraction pattern*. Similar to
the two-slit interference pattern, the central area is the brightest
and is surrounded by alternating dark and much less intense
brighter areas. Clearly the separation between these areas is
characteristic of the wavelength of the light falling on the slit and
a measurement of this separation can, in principle, be used to

measure the wavelength of the light. With a single slit the bright and dark areas on the screen are very broad and fuzzy and it is impossible to make a precise measurement, but if there are many slits, as in what is referred to as a *diffraction grating*, the interference effects are such as to produce very narrow bright lines which enable immensely precise measurements of wavelength to be made. Diffraction gratings are made, for example, by ruling many parallel lines with a diamond point on a glass surface.

4.7 Moving Forward

This last chapter has discussed the essentials of electromagnetic phenomena—electric and magnetic forces, electric currents and the relationship between magnetism, electricity and electromagnetic radiation. The ideas in this chapter and, indeed, in previous chapters were current towards the end of the last century and are often referred to collectively as 'classical physics'. However, to proceed further, for example, to consider the electromagnetic properties of materials, it is necessary to take on board some of the major and revolutionary advances in physics which took place in the first decades of the 20th century. In particular we have now to consider the detailed structure of atoms and the ideas of quantum mechanics.

CHAPTER 5

QUANTUM PHYSICS AND THE ATOM

Quantum Ideas, Nuclei, Electrons and Atomic Structure

5.1 Atomic Constituents—Electrons and Nuclei

In the last chapter the idea was introduced that an atom consists of negatively charged electrons and a positively charged nucleus such that the total negative charge of the electrons exactly balanced the positive charge of the nucleus. It is now time to explain how this idea came about. During the last half of the 19th century physicists studied the way in which electric currents flowed through gases. A gas is normally an insulator but, at a sufficiently low pressure and with a sufficiently high voltage (potential difference) between a positive and a negative electrode, a current will flow and the gas becomes luminous. Experiments of this sort are carried out in glass tubes, known as discharge tubes (see figure 5.1), and present-day 'neon tubes' are of this kind. At very low pressures the emitted light normally observed disappears but, if a hole is made, as shown, in the positive electrode, a glow is observed at the end of the tube. This glow can be attributed to a beam of negatively charged particles, which we now know to be electrons, being attracted from the negative towards the positive electrode. In the arrangement shown in figure 5.1 the electrons

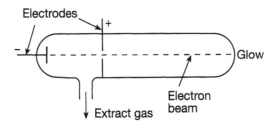

Figure 5.1: *The flow of electrons through a discharge tube.*

pass through a hole in the positive electrode, and finally strike the end of the tube. Where they strike the tube a glow is produced and the process is familiar in every home as being essentially responsible for television pictures. In the 1890s J J Thomson studied this process very carefully. He applied electric and magnetic fields to the narrow beam of electrons defined by the aperture and measured how the beam moved in relation to the strength of these fields. From these measurements he was able to deduce the ratio of the charge to the mass of the electrons.

If electrons are atomic constituents they can be knocked out of atoms. An atom, losing an electron, is then left with a net positive charge equal in magnitude to that of the electron. Such an entity is referred to as a *positive ion*. Ions can be produced in various ways, for example by knocking electrons out of an atom using x-rays. In addition, *negative ions* can also be formed when an extra electron attaches itself to an atom. Such ions have the property that, in a very damp atmosphere saturated with water, droplets can form on them, the water being attracted by the ionic charge. By studying the way such droplets move in an electric field it is possible to estimate the electric charge carried by a droplet, that is, the magnitude of the charge of an electron. This approach was developed particularly by R A Millikan who, in 1911, using oil drops rather than water drops, obtained an accurate value for the magnitude of the electron charge which is now known to be $e = 1.602 \times 10^{-19}$ C. Knowing the charge, Thomson's result enables the mass of the electron to be deduced. This turns out to be 9.11×10^{-31} kg. The important thing to note at this stage is that the electron mass is some 1836 times *smaller* than that of the lightest atom—hydrogen. So, in the light of this, we now have to consider where the rest of the atomic mass is located and also how the positive charge is distributed.

We have seen in section 3.2 that atoms in a solid are separated by distances of around 10^{-10} m. In a solid the atoms are squashed together and so one can (correctly) presume that atomic sizes are of this order of magnitude. In considering the way in which atoms are constructed, Thomson initially thought that the positive charge and the bulk of the mass occupied a sphere of atomic size and that the electrons were dotted around in this sphere like currants in a plum pudding (see figure 5.2(a)). However, developments in the next few years showed that this idea was completely wrong.

At the same time as Thomson was studying electrons, Becquerel and Pierre and Marie Curie in France and Rutherford in Britain were investigating the new phenomenon of *radioactivity* in which different forms of radiation were emitted by heavy atoms such as radium and uranium. These radiations will be discussed in section 8.6. Suffice it to say here that one of them, referred to as alpha-radiation, was found to consist of a stream of positively charged helium atoms (i.e. helium ions). In order to understand the details of atomic structure Rutherford suggested to two of his colleagues, Geiger and Marsden, that they should study the way these 'alpha-particles' were scattered by the electrical force due to the positive and negative charges in an atom when they passed through a thin sheet of material. This would test whether the 'plum pudding' idea was correct because, if it were, the expectation was that the alpha-particles would hardly deviate as they passed through the material.

This experiment was carried out using a thin gold foil and studying the direction in which the alpha-particles were scattered by the flashes they produced when they struck a zinc sulphide screen (see figure 5.2(b)). The flashes were due to sudden movements of electrons when 'struck' by alpha-particles.

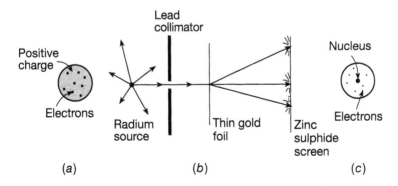

Figure 5.2: *(a) The 'plum pudding' atom; (b) the Geiger–Marsden experiment; (c) the nuclear atom.*

Although most alpha-particles were hardly scattered, a few were scattered through wide angles and even, very occasionally, bounced completely backwards. Rutherford was amazed at this latter observation and likened the result to shooting a 15 in shell at a piece of tissue paper and having it coming back and hitting you! He finally concluded in 1911 that all the positive charge and the bulk of the atomic mass must be concentrated in a very small volume—the *atomic nucleus*—at the centre of an atom as in figure 5.2(c) and that the 'bouncing back' effect occurred when an alpha-particle experienced a 'head-on' collision with a nucleus. Detailed analysis of the scattering indicated that the size of the gold nucleus, was of the order of 10^{-14} m, some four orders of magnitude smaller than the atomic size.

Thus was established the idea of the nuclear atom in which all the positive charge and the bulk of the atomic mass is concentrated in a nucleus, whilst the negatively charged electrons surrounded the nucleus, occupying a much larger volume. The next question is, of course, what are the electrons doing in this space? This can only be answered in terms of the ideas of quantum mechanics which were emerging over this same period.

5.2 The Rise of Quantum Mechanics

The first decades of the 20th century were clearly witnessing a revolution in the way physicists thought about the structure and behaviour of matter. Not least was the realization that the behaviour of systems with very low energy, such as individual atoms, was not described satisfactorily by the 'classical' mechanics embodied in Newton's Laws of Motion (section 2.2). The first indication of this came when attempts were made to understand the nature of the electromagnetic radiation discussed in the last chapter when emitted from a heated body due to oscillations of its component electrons. If the temperature is sufficiently high then we can actually *see* the radiation as visible light as, for example, emitted by the filament in an electric light bulb. The light emitted in this case is essentially 'white' and, as discussed in section 4.5, there is a spread in the frequencies emitted. With a bulb, radiant heat is also generated so that infrared rays of lower frequency than visible light are also present. The intensity of the radiation at the different frequencies is characteristic of the temperature of a heated body. For example, at a lower temperature, as with a domestic radiator, no visible light is present at all but there is still a characteristic spread of lower frequencies in the infrared region. An object at higher temperatures will emit visible light and, at extremely high temperatures, most of the radiation will be at frequencies characteristic of ultraviolet light and x-rays. There is a simple relationship (Wien's Law) between the frequency around which most of the radiation is emitted and the absolute temperature (T), namely

$$\text{frequency} = (10^{11} T) \text{ oscillations per second.}$$

Classical ideas were unable to account for the nature of these temperature dependent frequency distributions. In particular classical theory predicted that, whatever the temperature, the radiation would be increasingly intense the higher its frequency. This meant that even a cool body would be emitting light, ultraviolet rays and x-rays in dramatic contradiction to observation. The solution to this problem was first intimated by Max Planck in 1900, who suggested that the least energy that an oscillating system could have, other than no energy at all, was proportional to its frequency. This meant that it would be impossible to set very high-frequency oscillations in train because they would require much higher energy than would be available due to thermal motion. This would be particularly so when the emitting body was at a low temperature. The constant of proportionality, which is now called *Planck's constant,* is one of the key fundamental constants of physics. Planck's proposal thus has the form

Lowest energy of an oscillator = Planck's constant $\times$ frequency = $h\nu$

where h is the symbol universally used for the constant and has the value $h = 6.626 \times 10^{-34}$ J s. The Greek letter ν is generally used as the symbol for frequency. Planck went further than this, however, and proposed that the only allowed energies an oscillator could have—its *energy* levels—were integer multiples of this lowest energy (i.e. $n h \nu$ where $n = 0, 1, 2 \ldots$). So, whilst classically an oscillator's energy can creep up continuously.from zero (when it is at rest) to any desired value, with the Planck hypothesis the energy of an oscillator is *quantized*; only discrete values are allowed determined by the *quantum number n*. Suffice it to say here that this hypothesis accounted perfectly for the nature of the electromagnetic radiation emitted by a hot body, but it was completely contrary to the current classical theories and was the beginning of a major revolution in physics.

The next key development was connected with the *photoelectric effect*. This is the effect in which electrons (referred to in this context as *photoelectrons*) are emitted from a metallic surface when electromagnetic radiation, in the form of, for example, light or ultraviolet radiation, falls on to it. This can be understood simply as energy being carried by the oscillating electric and magnetic fields of the radiation being transferred to atomic electrons and knocking them out of the atoms. When this process was carefully studied three curious features emerged.

1. Contrary to natural expectations, the energy of the individual emitted electrons did *not* depend on the *intensity* of the radiation, although the more intense it was the more electrons were emitted.

2. The energy of the electrons depended directly on the *frequency* of the radiation.

3. No electrons were emitted if the frequency fell below a certain value.

These features could not be understood at all in terms of classical ideas which, for example, would suggest that the more intense the radiation, the more energetic its oscillations and, therefore, the more energetic the photoelectrons. It raised the question as to why weak blue light produces more energetic photoelectrons than strong red light. The solution to this dilemma was suggested by Einstein who, influenced by Planck's quantum ideas, suggested that in the photoelectric process radiation behaves like a stream of energy packages or particles having an energy proportional to the frequency of the radiation. These particles are called *photons* or *quanta* and the energy of each photon is given by photon energy = Planck's constant $\times$ frequency = $h\nu$.

In the photoelectric effect the photon gives up its energy to an electron. The more intense the radiation the more photons there are; hence more photoelectrons are emitted. The energy of a photoelectron is simply the energy of the photon, which depends only on the frequency, less the energy needed to escape from the metal. Clearly, below a certain frequency, there will not be sufficient energy to escape. Thus the three features of the photoelectric effect mentioned earlier are all neatly explained and Einstein was awarded the Nobel prize for this work—not for relativity!

This explanation does, of course, introduce a completely new idea into physics, namely that electromagnetic radiation which in many circumstances behaves as a wave—witness the discussion in section 4.4—can also behave as a stream of particles. This is the first instance of what is called *wave–particle duality* and is a key feature of the quantum revolution. The particle nature of electromagnetic radiation was subsequently directly confirmed by Compton in 1923, who showed that electrons scatter x-rays in just the same way as one billiard ball scatters another. This scattering could not be understood at all treating the x-rays as waves.

These ideas then stimulated Niels Bohr in 1913 to suggest a simple form for the structure of atoms. He proposed that electrons moved around the atomic nucleus in circular orbits, rather like the planetary system, constrained by the electric attractive force due to the positively charged nucleus. Classically, such electrons would emit electromagnetic radiation (see section 4.4) since they are continually accelerating—their direction of motion is changing all the time. They would thus continually lose energy and eventually spiral into the nucleus. To avoid this problem Bohr proposed that only certain orbits were allowed for an atomic electron and that, whilst in these orbits, no radiation was emitted. He specified these orbits by hypothesizing that in them the angular momentum (see section 2.3) of the electron is an integer multiple of $h/2\pi$ which is universally symbolized by $\hbar$. In mathematical terms the angular momentum is only allowed to have the values $n\hbar$ where $n = 1, 2, 3, \ldots$. This, in turn, means that the orbiting electron can only have certain energies determined by the value of the quantum number *n*. Bohr then proposed that radiation is only emitted when an electron jumps from a higher energy orbit to one of lower energy and that its frequency, v, is given by simply equating the energy released to hv, here using the idea of Einstein and treating the emitted radiation as a photon. This means that emitted radiation can only have certain frequencies, determined by the values of the quantum number *n* before and after the jump. Further, the energy of an electron can never be zero since the lowest value allowed for *n* is 1 (*not* 0) which means that an atomic electron is always in orbit and, therefore, does not fall into the nucleus.

It was a great triumph for Bohr that observed frequencies of the electromagnetic radiation emitted (the *atomic spectrum*) by hydrogen, which has only one electron, when bombarded by electrons in a discharge tube, agreed very closely with this simple theory. Such a spectrum is observed as a series of discrete lines, corresponding to the different allowed frequencies, when refracted through, for example, a glass prism. It was also very satisfactory that the radii of electron orbits calculated on this basis were consistent with the known sizes of atoms.

Bohr's theory, succesful as it was, could not be carried through to account for the more detailed properties of atoms. It was a hybrid theory relying on both classical and quantum ideas. For example the concept of an electron in orbit is classical and, as mentioned above, it should be continually emitting electromagnetic radiation. Yet, according to Bohr, radiation is only emitted when an electron jumps from one orbit to another and then it had to be treated as a photon rather than a classical wave motion. For this and other reasons the theory had to be developed further into a full quantum theory. This took place during the following decade and was based on the implications of wave–particle duality already emphasized.

5.3 Waves and Particles

As we have seen in the last section, there is good evidence that electromagnetic radiation, which, classically, is regarded as a wave motion, can also behave as a stream of particles—photons. The obvious question to ask is whether particles, such as electrons, can in turn exhibit wavelike properties. That this might be so was first conjectured by de Broglie in 1924. He noted that for a photon the energy is proportional to the frequency and, therefore, inversely proportional to the wavelength of the associated wave (see section 2.6). Then, using a simple relationship between energy (E) and the magnitude of the momentum (p) arising in relativity theory (E = pc where c is the speed of light—see section 7.5), he was able to deduce* a relation between the wavelength and the magnitude of the photon momentum, namely

$$\text{wavelength} = \frac{h}{p}.$$

He suggested that this relationship might also hold for material particles as well as for photons, which are, of course, massless. It was important to test his idea experimentally and this was done for electrons by Davisson and Germer in 1927. One characteristic of waves is that they can be diffracted (see section 4.6) and they found (interestingly by accident!) that electrons accelerated through a potential difference and then striking a nickel crystal were scattered preferentially in certain directions. This indicated that diffraction was taking place and clearly the regular lines of atoms in the crystal were playing the role of the lines which are engraved on a diffraction grating used to diffract light. The spacing between the lines of atoms was known from the diffraction of x-rays by the crystal and, knowing this, the wavelength of the 'electron wave' could be worked out. It agreed perfectly with de Broglie's suggestion. Similar confirmatory experiments have now been carried out with other particles and even atoms.

Thus wave–particle duality was firmly established. However a fundamental question remained: what is the nature of the wave associated with a material particle? Waves encountered so far have referred to some physically measurable quantity such as the

---

* The energy of a photon is $h\nu$, which we now equate to pc ($h\nu = pc$). Its wavelength is given by c/ν (see section 2.6) which, from the foregoing expression, is equal to h/p.

height of water, the density of air or the strength of electric and magnetic fields. What does the intensity of a 'matter wave' measure? The currently accepted interpretation was first suggested by Max Born following on detailed mathematical formulation of these quantum ideas by Heisenberg and Schrödinger. He interpreted the wave as carrying information about the location of the particle: its intensity at any point is simply a measure of the probability of finding the associated particle at that point. The particle is most likely to be found at positions where the wave is strong and least likely to be found where the wave is weak. Therefore the diffraction pattern of the wave associated with the electrons in the Davisson and Germer experiment had the equivalent of bright and dark areas obtained with the diffraction of light, but the 'bright' areas in the electron case are those where many electrons are detected, since the wave is intense there, and the 'dark' areas are those where few electrons are detected.

This interpretation brings with it profound conceptual problems. Consider the example of a particle moving along in a straight line with a definite momentum. From the above relationship the associated wave has a definite wavelength. Such a wave, representing a moving particle, must be a travelling wave (see section 2.6) and therefore extends endlessly in space. This means that, although we know the momentum of the particle precisely we have no idea where it is since the wave telling us about the probability of finding it somewhere extends to infinity.

Of course, in principle, we can know where a particle is by shining light on it and 'looking' at it. However, light is spread out—it is a wave motion—and so the best that could be done would be to localize the particle within one of the light's wavelengths. This means that there is an uncertainty in its position of the order of one wavelength. Further, the light carries momentum equal to h divided by this wavelength as mentioned earlier, so that in bouncing off the particle momentum of that order of magnitude is given to the electron. There is, therefore, also an uncertainty of that amount in the momentum of the particle. To be more certain of the particle's position would require a smaller wavelength which, in turn, would lead to a larger uncertainty in the momentum. Given that the uncertainty in the particle's position is simply the wavelength of the light and that the uncertainty in its momentum is h divided by this wavelength, it follows at once that

$$(\text{position uncertainty}) \times (\text{momentum uncertainty}) \cong h$$

where, the reader is reminded, the symbol $\cong$ signifies 'approximately equal'. It is used because the argument leading to this has been semiquantitative.

The above expression is a statement of what is known as *Heisenberg's Uncertainty Relation*. Clearly, as the momentum uncertainty approaches zero, then the position uncertainty approaches infinity (h divided by zero) as deduced earlier. Roughly speaking, if you know precisely where a particle is you cannot know what it is doing or, if you know precisely what it is doing, you cannot know where it is! Because of the small size of h this effect is also very small and only becomes significant at the atomic level; it has no importance in, for example, playing snooker. Indeed for entities much larger and heavier than atoms, such as snooker balls, quantum mechanics gives essentially the same results as classical Newtonian mechanics. Niels Bohr called this relationship between quantum and classical mechanics *the correspondence principle*.

It must be stressed here that the relation does not reflect some clumsiness in our procedures for measuring things; it is intrinsic to the particle–wave duality exhibited by material particles and electromagnetic radiation. It poses problems of interpretation and is still the subject of much debate. Consider, for example, the interference of light passing through two slits discussed in section 4.6 but treating the light as photons rather than waves. For a stream of photons the situation is straightforward in that the intensity of the waves relate simply to the probability of a photon being there. Thus there are many photons at the bright points of the interference pattern and few at the darker points. However consider just one of these photons arriving at the two slits. It can only go through one, so how does it 'know' that the other one is there and therefore participate in forming the interference pattern? The same question would arise if the experiment studied interference using a beam of electrons. There is no question of the photon or electron 'splitting up' and going through both slits; rather the associated wave goes through both slits and its intensity then determines the probability of the particle going through one or the other slit. This and many other similar questions have been written about extensively in much popular scientific literature.

5.4 Using Quantum Mechanics

The foregoing discussion has underlined the importance of using the wavelike properties of particles to understand certain physical phenomena. Clearly this approach had to be encompassed more formally in a theory which would enable the form and implications of these waves to be dealt with mathematically in just the same way that classical theory was encompassed in Newton's Laws of Motion (Chapter 2) and electromagnetic theory (Chapter 4). As mentioned earlier, the key physicists responsible for this development were Heisenberg and Schrödinger. The latter developed an equation—the *Schrödinger Equation*—which enabled the wavelike nature of a physical system to be described mathematically by a quantity called the *wavefunction* of the system. Associated physical quantities such as energy could also be deduced from the equation. A key element was that the wavefunction should make

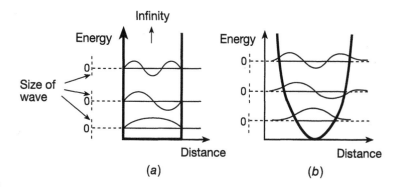

Figure 5.3: (a) Wavefunctions and energies for a particle in an infinite potential well. (b) Oscillator energies.

physical sense. For example it should only have one value at any point, otherwise an unacceptable ambiguity would be introduced into physics; if it had two values it would be unclear which one should be used to determine the probability of finding the associated particle at that point. This means that the wavefunction has to be continuous; it cannot jump from one value to another at the same point. It is requirements of this sort that lead to the quantization of energies in a system first envisaged for an oscillator by Planck and discussed in section 5.2.

The way in which this can happen can be illustrated most simply for a particle confined in a potential well with infinite sides. In Chapter 2 realistic potential wells were discussed. An infinite well is rather artificial but is much easier to deal with. Classically the particle in such a well can, in principle, have any energy from zero upwards; this is not the case with quantum mechanics. Since the particle is confined to the well (it has infinite sides), and cannot possibly escape, the wavefunction outside must be zero—there is zero probability of finding the particle there. This means that, if the wave function is continuous, the part *inside* the well must be zero at the two edges so as to join, without jumping, the zero wavefunction outside. A few different possibilities of arranging this are shown in figure 5(a) and it will be recognized that this is completely equivalent in mathematical terms to the different standing waves that can be set up on a vibrating string fixed at both ends (compare figure 2.8). The longest wavelength (corresponding to the lowest momentum and energy) possible is for the lowest wavefunction in the diagram. But this lowest energy is not zero: zero energy would require zero momentum and, therefore, infinite wavelength. This means that quantum mechanics does not allow a particle to sit at the bottom of the well with no energy at all. This minimum energy that a confined particle must have is referred to as *zero-point energy*. A few possible standing waves having increasingly shorter wavelengths and corresponding to higher allowed energies are shown in the diagram. We have here an example of quantized energies—only specific values are allowed. These referred to as *energy levels* and are analogous to the different harmonics mentioned in section 2.6 for a vibrating string. Roughly speaking, the more 'wiggly' a wave function the higher the associated energy.

Consider now the more complicated (mathematically) situation of a particle which classically can oscillate with frequency v. The potential well is identical in shape with that for a pendulum bob (see figure 2.4) and is reproduced in figure 5.3(b). Again the requirement that the wavefunction should be 'well behaved' limits the allowed energies, the lowest ones being shown in the diagram. It will be noted that the energy levels are separated by hv as suggested by Planck and are given by the simple formula

$$E_n = (n + \tfrac{1}{2})hv$$

where n can take the values 0, 1, 2 . . . .. The quantity n is known as a *quantum number* and such numbers occur in all quantum formulations giving the allowed energies of a system. It will be noted here that the lowest energy of an oscillator is $hv/2$ and not zero as hypothesized by Planck. This energy, like the zero-point energy for the infinite potential well, reflects the uncertainty relation. If the particle were at the bottom of the oscillator potential well with zero energy (i.e. zero momentum) there would be no uncertainty in its position or its momentum, contrary to the relation.

The shape of the wavefunctions corresponding to the different oscillator energy levels are also shown in figure 5.3(b) and it can be seen that they extend *beyond* the potential energy curve. This means that there is a finite probability of the particle being in a region which is not allowed classically. This has the implication that particles can penetrate barriers and can escape from situations where, classically, they would be confined. Consider, for example, a particle confined by two potential barriers (a sort of one-dimensional box) as shown in figure 5.4. The particle energy is lower than the height of the barrier and so classically it cannot escape.

However, as the diagam indicates, penetration of the wavefunction through the potential barrier takes place and this means that there is a finite probability of finding the particle outside the box. This, of course, is a quantum effect and is so small that it has no bearing on the use of boxes in everyday life! However it is of importance at the atomic and nuclear level, for example in understanding aspects of alpha-radioactivity in which alpha-particles escape from the nucleus (see sections 5.1 and 8.6).

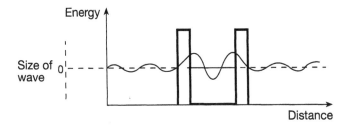

Figure 5.4: *A particle in a one-dimensional box.*

5.5 Atomic Structure

We can now extend this quantum approach and consider the way in which electrons are confined in an atom. Such electrons move in a potential well created by the electric field due to the positively charged atomic nucleus. The problem is now three dimensional as compared to the simple one-dimensional situations we have considered so far and the associated mathematical solutions of the Schrödinger equation (i.e. the wavefunctions) are significantly more complicated, but the same type of result emerges, namely that, for well behaved wavefunctions, only certain energy levels for an electron are allowed. The wavefunctions are now characterized by *three* quantum numbers, reflecting the fact that we are dealing with three dimensions. Considering a single electron, the allowed energy levels are precisely those conjectured by Bohr, discussed in section 5.2, except that the quantum number n does *not* specify the angular momentum of the electron: it gives information about the shape, not least the 'wiggliness', of the wavefunction. The electron orbital angular momentum is, nevertheless, specified and its value and direction are given by two other quantum numbers (denoted by l and m) whose values, for a given n, are limited. Also, as Bohr conjectured, it is measured in units of $\hbar\ (= h/2\pi)$.

The situation for a single electron is simple and the energy levels and wavefunctions of the hydrogen atom, which has only one electron, can be calculated very straightforwardly. The allowed frequencies of the electromagnetic radiation (referred to as *spectral lines* in the hydrogen spectrum) when the electron jumps from one level to another can therefore be readily calculated by equating the energy difference between the two levels to $h\nu$, the energy of the emitted photon. Experimental tests showed that, although there was fairly good agreement, some of the lines appeared to consist of *two* closely separated lines. Similar anomalies appeared in the spectrum emitted when the emitting hydrogen atoms were placed in a magnetic field. To account for these discrepancies, it was proposed in 1925 by Goudsmit and Uhlenbeck that the electron itself had an intrinsic angular momentum and was always spinning like a top. The nature of the anomalies could be accounted for if the electron spin had the value $\hbar/2$ and we speak of the electron as having spin $\frac{1}{2}$. It also followed that, since the electron carries an electric charge, its rotation would lead to an electric current and, in turn, to the electron behaving as a miniscule magnet. The splitting of the spectral lines could then be understood in terms of small changes in magnetic energy depending on which way the electron spin was pointing in relation to the magnetic field produced by the orbital motion of the electron. Similar explanations also accounted for the behaviour of atoms in external magnetic fields.

For atoms containing two or more electrons the situation becomes much more complicated, because not only is there the attractive electric force between the nucleus and the electrons but also there is electrical *repulsion* between the electrons themselves. This is, however, a complicated matter of detail. Another much more fundamental issue is consideration of which energy level each electron in a multi-electron atom occupies. The natural assumption is that in a stable atom every electron is in the same energy state— the lowest one possible; in a higher state an electron would be expected to drop to the lowest state, emitting radiation in the process. This would lead one to expect that all chemical elements, i.e. the different possible atoms, would be very similar. However, it is found that atoms differing even by one electron can have dramatically different chemical and physical properties.

The solution to this basic problem was put forward by Pauli, also in 1925, who hypothesized that all electons are *absolutely* identical (not like two billiard balls with their small imperfections) and that no two identical particles of spin $\frac{1}{2}$ can occupy the same quantum state; that is each electron in an atom must occupy a state with a different set of quantum numbers. This is known as the *Pauli exclusion principle*. It is now known to be correct and, as we shall see later, has a profound effect on the structure of all states of matter. In an atomic energy level whose energy is specified by the quantum number n (see above) only a certain number of electrons can be accommodated depending on how many different values of the quantum numbers l and m are available to it and in which direction the electron spin is pointing (colloquially referred to as 'up' or 'down'). For example, the lowest level can only accommodate two electrons; it is then full and we speak of a *filled shell*. Atoms such as this (helium in this simplest of cases) with only completely filled levels are very stable and hardly interact; they include what are known as the inert gases (helium, neon, argon etc). Moving to an atom having a nucleus with three units of positive charge and three electrons, two will be in the lowest energy level and the third is alone in the next level up. In some respects the resultant atom is rather like a hydrogen atom having a single electron outside a tight core consisting of the nucleus and the two inner electrons which together have unit net positive charge (three positive charges on the nucleus and two negative charges on the two electrons). This atom is lithium and other atoms with filled inner shells and one spare electron are sodium, potassium etc, all having similar properties. And so we could go on to atoms with higher and higher positive charge on the nucleus and having larger and larger numbers of electrons. These electrons fill higher and higher energy levels according to the limitation on their numbers required by the exclusion principle. In moving through these different atoms further similarities in atomic structure and associated properties are found. It is in this way that we understand the *periodic table* of atoms first introduced on an empirical basis by Mendeleev in 1869, which exhibits the recurrently similar properties of the different chemical elements.

5.6 Atomic Radiation

We have seen that radiation is emitted when an electron jumps from one energy level to another having lower energy. It may have been knocked into the higher level in a discharge tube by

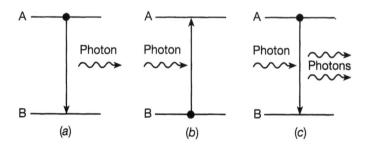

Figure 5.5: *(a) Spontaneous emission, (b) absorption and (c) stimulated emission of atomic radiation with* $h\nu = E_A - E_B$.

energetic electrons moving across the tube. The emission of radiation is illustrated in figure 5.5(a) for a transition between two levels labelled A and B having energies E_A and E_B. The associated frequency ν is related to the energy difference between the levels by the usual formula given in the diagram. This is referred to as spontaneous emission. For small energy differences the frequencies lie in the optical range. However sometimes the energy difference can be large. This happens, for example, in an x-ray tube when atoms are bombarded by high-energy electrons. These can knock out electrons from filled energy levels deep down in the atom leaving, as it were, a vacancy or hole which the exclusion principle allows to be filled. An electron from a much higher energy level can jump into the hole and the resultant radiation can then have a high frequency in the x-ray range (see table 4.1).

If an electron is in an energy level (e.g. A in figure 5.5) and has the possibility of making a transition to a lower level, the point (in time) at which it will make the transition cannot be predicted. All that quantum mechanics can tell us is that *on average* it will spend a certain time (referred to as its *mean life*) in that level. Typically, in an atomic transition, such lifetimes are of the order of 10^{-8} s. This is an uncertainty in the time at which an event happens and there is another important associated uncertainty relation of the form

$$(\text{time uncertainty}) \times (\text{energy uncertainty}) \cong h$$

which means that the emitted photons in a transition do not have the exact frequency given by the formula because of the (small) uncertainty in the energy of the level A relating to its lifetime. The lines in an atomic spectrum therefore have a small width in their frequency. It is by measuring this width and using the above uncertainty relation that the lifetime of atomic states can be estimated.

Another possible radiation process is absorption (see figure 5.5(b)) in which a photon of the right frequency excites an electron from the level B to the level A. This means that if light with a spread of frequencies passes through a gas the gas atoms will absorb light of certain frequencies when electrons are excited to higher levels. In turn this leads to dark lines in the spectrum of the light that has passed through the gas. This is called an *absorption spectrum* and is clearly characteristic of the gas. Absorption spectroscopy is an important analytical process and has been used, for example, to identify atoms (and molecules) in space through which sunlight and starlight has passed.

Stimulated emission of radiation can also occur when an electron in an atom is in an excited level, for example level A in figure 5.5(c), and radiation falls on it. If this radiation has the frequency appropriate for a transition to the level B, then it stimulates the electron to make that transition and so the end product is two photons. The important thing about this process is that the two waves associated with the photons are in phase (coherent) in that their crests and troughs coincide and they also travel in the same direction. This means that they reinforce each other very strongly. Of course, if the state A has a typical atomic lifetime then it will, in any case, make this transition spontaneously and stimulated emission will be hard to observe. However some special states (referred to as *metastable*) can have a very long lifetime (as long as 10^{-3} s) and there are ways of 'pumping' many atoms into such states. Passing light of the appropriate frequency through such a 'pumped-up' collection of atoms and reflecting it backwards and forwards then results in a great deal of stimulated emission and an immensely powerful resultant coherent light wave is produced. This is the principle underlying the operation of a *laser* where the word derives from the phrase Light Amplification by Stimulated Emission of Radiation.

5.7 Moving Forward

We have seen in this chapter that processes and structures at the atomic level can only be understood in terms of the radically new ideas of quantum theory. The classical approach breaks down and phenomena have associated with them a basic uncertainty relating to the specification of their precise details. Probability conditions the outcome of measurements in both space and time. Nevertheless using quantum theory it has been possible to understand the structure and behaviour of atoms and the way they emit and absorb electromagnetic radiation. Apart from basic quantum theory it has also been necessary to attribute spin to electrons and to set limits on the number of electrons that can occupy a particular quantum state through the requirements of the exclusion principle. Our next task is to see how these ideas affect the nature of matter itself, particularly its electromagnetic properties.

CHAPTER 6

PROPERTIES OF MATTER—SOME QUANTUM EXPLANATIONS

Mainly about the Behaviour of Electrons in Matter

6.1 The Origins of the Interatomic Force

The nature of the force between two atoms was discussed in section 2.1. Broadly speaking it is a force that is attractive when two atoms come close to each other and then eventually becomes repulsive when they become very close. In the last chapter the essential structure of an atom was outlined: a very small but massive positively charged nucleus surrounded by electrons in different quantum states. For a given atom, these states are filled with the requisite number of electrons according to the limitations of the Pauli exclusion principle. In an atom the innermost electrons are in filled 'shells' (i.e. in stable sets of quantum states, which, by the exclusion principle, cannot accommodate any more electrons) and usually there are a few electrons in unfilled outer shells. When two atoms approach each other electrical forces clearly come into operation and these, together with quantum effects, determine the detailed nature of the force between two atoms. The precise explanation of the force depends, as might be expected, on the details of the atomic structure and there are essentially three types of mechanism in operation. These will now be outlined.

Ionic Force. This force is the simplest to understand and accounts, for example, for the binding of sodium to chlorine in common salt. Sodium is an atom with 11 electrons, ten of which are in filled shells, and one is fairly loose. Chlorine, on the other hand, has 17 electrons and is just one short of having completely filled shells—it has a 'hole' in the shell. When these two atoms approach each other it is easy for the spare sodium electron to jump over into the hole and the resultant system is much more stable, but the sodium atom, having lost a negatively charged electron, has a net positive charge, whilst the chlorine atom, having gained an electron, has a net negative charge. We have, therefore, a positive sodium ion close to a negative chlorine ion and clearly there will be an attractive electric force between them due to their opposite charges and it is this which holds them together in the molecule (see figure 6.1(a)). If they move very close together the electrons in the filled shells start encroaching on each other's orbits. This is forbidden by the exclusion principle and so repulsion begins to take over. In addition there is also an increasingly important repulsive force due to the close approach of the positively charged atomic nuclei. Hence the general features of the interatomic force are produced. The joining together of two atoms by the ionic force is referred to as *ionic bonding* and manifests itself in many diatomic molecules.

Covalent Force. The ionic force occurs because an electron is handed over from one atom to the other. Covalent forces, on the other hand, result from the 'sharing' of electrons between each

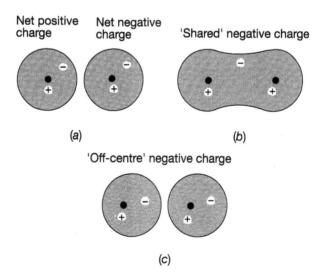

Net positive   Net negative
charge   charge   'Shared' negative charge

(a)   (b)

'Off-centre' negative charge

(c)

Figure 6.1: *Charge distributions relevant to interatomic forces: (a) ionic; (b) covalent; (c) Van der Waals.*

atom. To understand this the simplest system to consider is the hydrogen molecule, consisting of two hydrogen atoms held together by the covalent force. Each hydrogen atom has a single electron and, as the atoms approach each other, the electron of one begins to feel the electrical attraction due to the positive nucleus of the other. At a close enough distance the electrons move across between the atoms and partially occupy each other's quantum states—the electrons are shared between the two atoms. It will be recollected (see section 5.5) that the lowest state of the hydrogen atom can only accommodate two electrons (with opposite spins) and therefore, to some extent, the state is being filled. This makes for a very stable structure.

The two electrons tend to spend most of their time *between* the two nuclei (see figure 6.1(b)) and the resultant central negative charge draws the two positively charged nuclei together. Of course, repulsion begins to set in when the two nuclei become very close together because of their like charges. Covalent forces play a major role in the formation of more complex molecules and the joining together of atoms in this way is referred to as *covalent bonding*. When an atom has several electrons outside filled shells this, in turn, enables covalent bonds to be formed with several other atoms and this mechanism underlies the formation of a wide spread of molecular structures from water (two hydrogen atoms and one oxygen atom) through to the immense complexity of DNA.

Van der Waals Force. This is a much weaker force than the two just discussed and does not involve any transfer of electrons between the two interacting atoms. It is, for example, the origin of the force between inert gas atoms (atoms which only have filled shells of electrons). In such atoms the electrons move around in their quantum states and although on average the electron charge is distributed symmetrically about the nuclear charge it is sometimes 'off-centre'. For example in figure 6.1(c) the average negative electron charge in the left-hand atom is displaced slightly to the right of the nucleus. Since this charge is now nearer to the right-hand atom than the (positive) nuclear charge, there is a net pull on the nucleus of the right hand atom and a slight push on its electrons, leading to a corresponding off-centre distribution of charge in that atom. As a result of this lopsidedness of the two charge distributions there is small attractive force between the two atoms, but the force is not strong enough to hold such atoms together as a molecule and this is why they are referred to as *inert*. However, this force plays an important role in explaining intermolecular forces, particularly when the molecules under consideration have an *intrinsic* lopsidedness because they consist of two or more different types of atom.

It is in terms of these different interatomic forces that we understand and explain the properties of the many varieties of molecule and also the cohesion of atoms in the various forms of matter. As far as molecules and their behaviour are concerned this is largely the province of chemistry and little more will be said about them except to note that their 'bulk' motions are also quantized. By bulk motions is meant motions of their component atoms. For example, in a diatomic molecule (think of it as being like a dumb-bell) the whole molecule can rotate with angular momentum quantized in units of $\hbar$ and with corresponding quantized energy levels. Similarly the two atoms can vibrate about their equilibrium separation with resultant vibrational energy levels of the type encountered in section 5.4. As a result, in addition to electrons in molecules making transitions from one energy state to another as in atoms, there can also be transitions between the rotational and vibrational energy levels, all leading to the emission or absorption of electromagnetic radiation. Molecular spectra are, therefore, much more complicated in their structure than atomic spectra, particularly for molecules containing three or even more atoms.

In Chapter 3 various properties of the three states of matter were explained in terms of the interatomic and intermolecular forces just discussed. We now extend this discussion further and seek explanations and understanding of the electromagnetic properties of matter.

6.2 Conductors and Insulators

In section 4.2 it was mentioned that an electric current through a metal is due to a flow of electrons. Our task now is to understand how electrons, which are held in an atom by the electrical attraction of the nucleus are able, in a metal, to escape from this attractive force. The process can most simply be understood by considering the potential well in which the atomic electrons are confined. This is shown in figure 6.2 (a) for a single atom. The attractive force due to the nucleus becomes stronger the nearer the electron is to the nucleus and, correspondingly, the potential well becomes deeper. The further the electron is away from the nucleus the weaker the force and so the well gradually levels out at large distances. In the well the horizontal lines represent the quantized energy levels in which the electrons are distributed according to the exclusion principle. When the atoms are brought very close to each other to form a solid the electrons can feel the attraction not only of their own atomic nucleus but also, to some extent, that due to neighbouring nuclei. This increased attraction therefore lowers the potential energy curve between atoms as shown in figure 6.2(b) for a small row of atoms. As an electron moves away from its nucleus its potential energy increases but, before it gets to the brim of the well as in the case of an isolated atom it begins to feel the attraction of the neighbouring nucleus and the potential energy begins to drop again. This means that the levels A and B in which an electron would be confined in an isolated atom now extend throughout the solid; they sit above the top of the row of potential wells. Of course there are many levels A and B—one for each atom—and the result is a bundle of levels all with slightly different energies forming what is called a *band*.

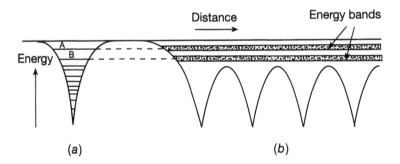

Figure 6.2: *Electron potential wells: (a) a free atom; (b) atoms in a solid.*

In figure 6.2(b) the bands originating in the levels A and B are shown. Sometimes such bands can be well separated, sometimes they are very close and sometimes they overlap.

Consider now a solid formed from atoms for which normally there is one electron in the level B. In the solid state this electron can now move around the solid; it is no longer constrained by the potential well as in an isolated atom. Of course, each atom will contribute an electron and so there is a very large number of electrons able to undergo this movement. Each one will be in a different state within the band, as required by the exclusion principle. If not all the levels in a band are occupied (see figure 6.3(a)) then, when an electric field is applied, electrons in the band can pick up small amounts of energy and move to unoccupied levels. This means that they can move under the influence of the field and, therefore, carry an electric current. Such electrons are referred to as *conduction electrons* and the band as a *conduction band*. We have here the situation which occurs in a metal where there is always a conduction band just partially filled with electrons. It should also be mentioned here that energy can also be given to conduction electrons by a heat input and they can then carry this heat energy through the solid. This is a much more effective way of conducting heat than the passing on of atomic vibrations as described in section 3.5. This is why metals are much better heat conductors than non-metals.

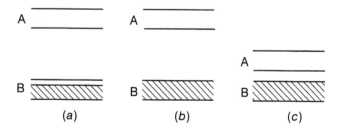

Figure 6.3: *Bands A and B in (a) a conductor (unoccupied levels in band B), (b) an insulator (band B full), (c) a semiconductor (small band gap).*

However, the electrons are not completely free to move. In most solids the arrangement of the component atoms is not absolutely regular; there are *imperfections*. There may also be *impurities*. In addition, the atoms are not stationary: they are oscillating in a random way about their equilibrium positions due to the thermal energy of the solid (see section 3.2). These different effects impede the motion of the electrons and lead to the phenomenon of *electric resistance*. As the temperature of a metal is lowered the thermal vibrations decrease in amplitude and so it is to be expected that its resistance will also decrease; this is indeed observed experimentally. Conversely, electrons moving through a solid under the influence of an electric field (an electric current) and colliding with atoms increase the amplitude of the atomic vibrations, leading to a corresponding rise in temperature. This accounts for the heat generated in the wires of, for example, electric fires and electric light bulbs.

Consider now an atom in which the level B has the full complement of electrons allowed by the exclusion principle. This means, in turn, that in the solid the band corresponding to the level B is full (see figure 6.3(b)). If an electric field is applied the only way that an electron can have more energy is for it to jump into the empty band corresponding to the level A. However, if there is a large gap between the two bands it will not have sufficient energy to do this and hence no change of electron energy can take place and no electric current can flow. This is the situation with electric *insulators* such as glass. Of course, if a massive electric field is applied then electrons can make the jump to the unoccupied band and current flows, leading to a breakdown of the insulator.

6.3 Semiconductors

In some solids—silicon and germanium are important examples—the situation arises where band B is full but band A is very close to it (see figure 6.3(c)). In terms of our discussion about insulators, such a solid should not carry an electric current since the band B is full. However, because band A is very close to band B, electrons from B can sometimes jump into A because of the energy they can acquire from thermal excitation. Clearly, the higher the temperature the higher the thermal energy of the electrons and the more likely they are to jump into band A. This band will not have the full complement of electrons allowed by the exclusion principle and therefore behaves as a conduction band so that an electric current can pass. Such a solid, because it can carry a current by this means, is referred to as a *semiconductor*. Its conductivity is much lower than that of a metal because there are far fewer electrons participating. The conductivity is also *very* dependent on temperature and a semiconductor resistance can decrease by a factor of around two for every ten degree temperature rise. Its conductivity can also be increased by shining light onto it, the photons in the light giving energy to electrons in band B, enabling them to jump into band A. Such a semiconductor is referred to as a *photoconductor*.

However, the most effective way of increasing the conductivity of a semiconductor is to 'dope' it with an impurity. Consider the semiconductor germanium with a small amount of arsenic as an impurity. The arsenic atom is very similar to germanium except that it has an additional electron. If a germanium atom is replaced by an arsenic atom in a germanium crystal then there is a spare electron. It turns out that the energy level of this electron is just below the bottom of the band A as shown in figure 6.4(a). Its energy level is so close to this band that, unless the temperature is very low indeed, thermal energy is enough to cause it to jump into the band and therfore act as a conduction electron. Impurities of a

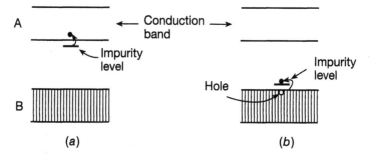

Figure 6.4: *Semiconductor band structures: (a) a donor impurity; (b) an acceptor impurity.*

few parts in a million are sufficient to provide a very large number of conduction electrons and can lead to a thousandfold increase in the conductivity. Since the impurity *donates* electrons it is known as a *donor impurity* and, because the current is carried by these (negatively charged) electrons, the doped germanium is referred to as an *n-type semiconductor*.

Another way to increase the conductivity of germanium is to introduce the impurity gallium. The gallium atom has one electron *less* than germanium and when introduced as an impurity there is an impurity energy level available for an electron just above the band B as shown in figure 6.4(b). This means that, because of its thermal energy, an electron from the band B can jump into this level leaving, as it were, a 'hole' in the band. When an electric field is applied another nearby electron in band B can jump into the hole and this process can continue so that the hole moves steadily through the germanium under the influence of the electric field. This is another form of electric current since electric charge is being displaced through the solid. An impurity of this form is referred to as an *acceptor* impurity since an electron from the band B is accepted into an empty energy level due to the impurity. Also since the hole travels in the opposite direction to the electrons it behaves as though it had a positive charge and we now speak of a *p-type semiconductor*.

Both n-type and p-type semiconductors play a profound role in the construction of all electronic apparatus from transistor radios, television and pocket calculators through to computer systems at all levels of scale. The basic ingredient is the *p-n junction* (referred to as a diode) consisting of p-type and n-type semiconductor layers in contact with each other as in figure 6.5(a). This can be arranged within the same crystal—usually silicon or germanium. It is found that with such an arrangement electrons can be drawn through it if the positive pole of the battery is connected to the p-layer and the negative pole to the n-layer. Roughly speaking, electrons are continually 'pumped' into the n-region and removed from the p-region, this latter process being equivalent to pumping 'holes' into that region. However, if the battery is connected the other way round electrons are 'sucked out' of the n-region and holes out of the p-region. The carriers of electricity are therefore

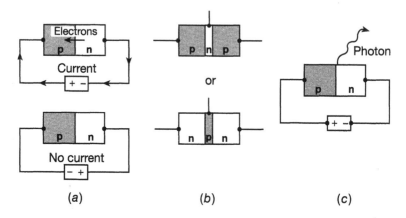

Figure 6.5: *Semiconductor devices: (a) a diode; (b) a transistor; (c) an LED.*

removed and the flow of electricity stops. Such an arrangement therefore allows an electric current to flow one way but not the other. A more complex arrangement is to have a n–p–n or a p–n–p junction with a thin middle region as in figure 6.5(b). This is known as a *transistor* (or a *triode*) and has the property that the electric potential on the middle section can be used to control the flow of current through the device. This enables, for example, a weak electrical signal being fed to the middle section to cause a large change in the current through the device and hence to amplify the signal.

As a current flows through a p–n junction electrons frequently fall into holes. Both obviously disappear and energy is released. In some materials this energy is sufficiently large to lead to the emission of a visible photon (see figure 6.5(c)). This is the basis of operation of LEDs (*light emitting diodes*) which are used in the visual displays of many electronic devices. Conversely, by shining light onto such a junction, electrons can be excited into the conduction band and these, together with the resultant holes, lead to flow of current if an external circuit is connected. This is the mechanism underlying the functioning of *solar cells* used for the generation of electricity, for example, in satellites.

Associated with the development of diodes, transistors etc are the techniques of miniaturization. These allow many semiconductor components of the above kind to be connected together in an extremely small space within the same crystal. It is possible to incorporate very complicated circuits inolving literally tens of thousands of such components on thin wafers of material with areas of the order of one square centimetre or less. These are referred to as *integrated circuits* and the wafer containing them as a *chip*, often called a *silicon chip*. That such complex circuits can be contained in such a small space has two benefits. First it has enabled the construction of very small but highly sophisticated electronic devices of which the pocket calculator and the digital watch are common examples. Second, because the chips are so small, signals between the different components can move very quickly, so enabling the development of high-speed large-scale computers.

6.4 Superconductivity

In section 6.2 the various causes of resistance to the flow of electrons in an electric current were mentioned—crystal imperfections, impurities and thermal vibrations of the component atoms. In a pure, perfect crystal at low temperature it would therefore be expected that the electrical resistance would be small and this is observed experimentally. However in 1911 Kammerlingh Onnes, a Dutch physicist, using liquid helium as a coolant (see section 6.5) found that below the extremely low temperature of $4.15\,\text{K}$ (i.e. 4.15 degrees above absolute zero—see section 3.3) the electrical resistance of mercury actually became zero rather than just small. This phenomenon was subsequently observed in many other substances at low temperatures. It manifests itself most dramatically if a current is set flowing in a closed loop of super-conducting wire, for example by moving a magnet near it (see section 4.3). Such a current should flow for ever since there is no resistance and there are, indeed, examples where currents have been kept flowing in such loops for years. This sudden drop of resistance to zero below a certain *critical temperature* cannot be accounted for in terms of the physical processes of conduction we have considered so far and heralds a new phenomenon referred to as *superconductivity*.

An understanding of what was happening was not forthcoming until 1957 when Bardeen, Cooper and Schrieffer proposed a theory which hinged on the fact that, when moving through a material, electrons experience not only a *repulsive* force between each other due to their negative charges but also an *attractive* force due to their interaction with the atoms in the material. This latter force arises because one electron moving near an atom displaces it slightly and that displacement, in turn, can have an effect on another electron. The net result is that the two electrons experience a small attraction to each other. At sufficiently low temperatures that the thermal vibrations of the atoms do not disturb the situation, this attractive force overcomes the repulsive force and results in electrons coordinating their motion in pairs. In the superconducting state the resistive scattering of one electron by an atom is exactly cancelled by the scattering of the other electron in the pair and there is no net effect. All the electrons move together in a coherent way and there is no resistance to their motion. Of course this is a *very* rough and ready account of a very sophisticated quantitative theory. What should be clear, however, is that the phenomenon of superconductivity is a quantum effect: no classical explanation is possible.

Another interesting effect observed with superconductors—the *Meissner effect*—is that if a magnetic field is applied it does not penetrate into it. The electrons flowing in the superconductor change their motion in such a way as to create an exactly opposite field which just cancels the applied field inside it. However, if this applied field is above a certain strength—the *critical field*—the cancellation cannot be sustained. The magnetic field penetrates the superconductor and it becomes a normal conductor.

In recent years it has been found that superconductivity in certain alloys can be achieved at much higher temperatures, of the order of 150 K, and the search is on for 'room-temperature' superconductivity which would enable large-scale loss-free transmission of electric power. Superconducting wire is, however, already used extensively in powerful electromagnets. For such magnets, large electric currents are needed and, in a superconductor, the current flows continuously without any heating and energy loss due to electrical resistance as with a normal electromagnet. The only expenditure is that required to keep the superconducting wire below its critical temperature.

6.5 Magnetism in Solids

We saw in section 4.3 that a magnetic field is created when an electric current—a flow of electric charge in the form of electrons—passes through a wire. It therefore follows that any electron in an atom possessing some angular momentum (see sections 5.2 and 5.5), and which can be pictured as orbiting about the atomic nucleus, will create a small magnetic field due to its rotation. In addition, electrons, having an intrinsic spin, also create a magnetic field and behave as miniscule magnets. Because of these two sources of a magnetic field in an atom it is to be expected that atoms themselves behave as small magnets. In general this is true except for those atoms in which there are no electrons outside filled shells. This is because in such atoms, for every electron in a filled shell rotating or spinning in one direction, there is another rotating or spinning in the opposite direction. The corresponding magnetic fields then exactly cancel each other and there is no resultant field. This means that in all atoms their associated magnetic fields derive *only* from those loose electrons outside filled shells; there is no contribution from the filled shells themselves. Solids whose atoms behave as magnets will have magnetic properties and these manifest themselves in two forms—*paramagnetism* and *ferromagnetism*.

Paramagnetism. In general, in a solid whose atoms behave as magnets, the magnets are pointing randomly in all directions and there is no resultant magnetic field. However, if an external magnetic field is applied then, just as a magnetic compass needle lines up with the earth's magnetic field, so it would be expected that the magnetic atoms in the solid will line up and all point in the direction of this field. This does happen to some extent and the result is that the solid becomes *magnetized* in such a way that the magnetic field due to the solid *enhances* the applied magnetic field. However the 'lining up' is not perfect. A compass needle is generally not subject to any disturbance. On the other hand, the atoms in a solid are being continually disturbed by thermal vibrations. These, of course, become more violent the higher the temperature and it is therefore to be expected that the 'lining-up' effect will reduce as the temperature is increased. Experimental observation confirms this expectation for non-metals. For metals the situation is somewhat different and more complicated because of the presence of conduction electrons (see section 6.2). This enhancement of an applied magnetic field, which occurs in many substances, is known as paramagnetism.

Ferromagnetism. In five elements, of which iron is the best known, it is found that the atomic magnetization results mainly from the spins of electrons in unfilled shells (see section 5.5). In the atoms of these elements the most stable configuration—the configuration with lowest energy—is one in which the spins of these electrons are mostly parallel to each other. This arises because with parallel spins the exclusion principle (section 5.5) requires each of the electrons to be in a different quantum state. The motion of each electron is therefore different so that they tend to keep well apart from each other. This, in turn, means that they have lower energy in this sort of configuration since, being mainly well separated, they experience less repulsion due to their electric charges. Naturally, the atom generally stays in this lowest-energy state. This effect continues to operate when the atoms are assembled in a piece of metal and, indeed, operates between nearby atoms and within the band structure. As a result the spins of the different electrons 'line up' *without* the imposition of an external magnetic field. This means that the magnetic fields of the 'lined-up' electrons all point in the same direction, leading to a strong resultant field. Materials which behave in this way are said to be ferromagnetic. Again, as with paramagnetism, this lining up cannot be sustained at high temperatures because of thermal agitation and above a certain temperature—known as the *Curie temperature* (1043 K for iron)—it does not occur.

Further, this lining up only happens within small *domains* of the metal having sizes in the range one-tenth of a millimetre to a few millimetres. Within a domain the lining up is perfect but the different domains in a ferromagnetic material will generally have their directions of magnetization pointing in random directions as illustrated symbolically in figure 6.6(a). If they lined up

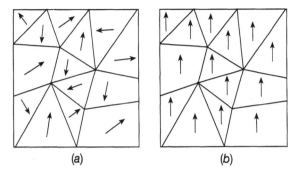

Figure 6.6: *Domains in a ferromagnetic material: (a) unmagnetized, (b) magnetized.*

throughout the material this would be a state of high energy because of the very large magnetic field caused by the domains all pointing in the same direction, and the system naturally tends to opt for lower-energy states. However, if the material is placed in an external magnetic field, then those domains pointing in the same direction as the field tend to grow and others rotate their direction of magnetization in the direction of the field (see figure 6.6(b)). The additional field due to the ferromagnet, which, of course, enhances the applied field, can be very strong indeed. If the applied field is then removed most of the domains will stay in position and we then have the well known *permanent magnet*. This retained memory by the material of the direction of the original applied field is referred to as *hysteresis*. However this configuration is somewhat unstable and, for example, hitting a magnet with a hammer can reduce its magnetism—the domains revert back to random orientations and lower energy.

Diamagnetism. Even if an atom does not behave as a small magnet, because all the electrons are in filled shells, there is still a small magnetic effect when a material formed from such atoms is placed in a magnetic field. In these atoms electrons in filled shells are moving although, as pointed out earlier in this section, their orbital rotations and the corresponding magnetic effects exactly cancel. However, because they are moving and carry an electric charge, when a magnetic field is applied they experience a force (see section 4.3). This force modifies the motion of each electron with the consequence that the orbital rotations do not exactly cancel out. There is a net resultant angular momentum and this leads to a small magnetic effect. It turns out that the magnetic field due to this induced magnetism is always in the *opposite* direction to the inducing field and therefore *reduces* its effective size. Diamagnetism then acts in the opposite way to paramagnetism and ferromagnetism, which enhance the applied magnetic field. It will, of course, also arise in the filled shells of atoms with loose electrons but its effect is relatively very small compared with the magnetism due to the loose electrons and can generally be ignored. Unlike paramagnetism and ferromagnetism, it is also independent of temperature since thermal vibrations do not affect the motion of electrons within the atom, only the motion of the atoms as a whole.

6.6 Superfluidity

In section 6.4 there was discussion of superconductivity in which electrons can flow through a metal and experience no resistance to their flow when the metal is cooled to a very low temperature. This, as was stressed, is a quantum effect and here we discuss a similar, although not directly related, phenomenon which takes place in liquid helium. The nature of the helium atom was mentioned in section 5.5. It consists of a nucleus together with two electrons in a filled shell. Since the shell is filled their total angular momentum is zero. The nucleus also has zero angular momentum and so, therefore, does the atom as a whole—it can be regarded as a particle of spin 0. Entities whose spin is an *integer* (0, 1, 2, . . .) multiple of $\hbar$ are referred to collectively as *bosons* (named after Bose, an Indian physicist). They are not subject to the Pauli exclusion principle (see section 5.5) unlike *fermions* (named after Fermi, an Italian physicist) such as electrons which have spin equal to a *half-integer* ($\frac{1}{2}, \frac{3}{2}, . . .$) multiple of $\hbar$. This means that an assembly of helium atoms, having spin 0, can in principle *all* occupy the same energy state.

At very low temperatures, when thermal agitation is small, a gas of helium atoms will condense into a liquid. This happens at around 4.2 K and, curiously, the helium will remain as a liquid down to the lowest possible temperature—in principle, absolute zero (0 K). The only way it can be solidified is to increase the pressure on it very significantly; roughly speaking to squash the atoms together. However, as it is cooled below the above liquefaction temperature a remarkable phenomenon occurs. An effective way to cool it is to pump away the helium vapour created by evaporation, which lies above the liquid helium. As it is pumped away more of the liquid evaporates and this requires energy, which is taken out of the remaining liquid, so reducing thermal agitation and, in turn, the temperature. The evaporation process is accompanied by the usual phenomenon of boiling (bubble formation) but when the temperature reaches 2.18 K the boiling suddenly stops although the evaporation process still continues. The physical interpretation is that suddenly the liquid is able to conduct heat without any resistance at all so that the heat no longer remains localized leading to the formation of bubbles. Coupled with this it is found that the liquid is able to flow down a very thin tube without experiencing any resistance to its flow. It also exhibits the remarkable behaviour of being able to climb quite quickly up the walls and out of its container as a thin film!

The liquid helium has suddenly changed its nature in some way and part of it has become what is called a *superfluid*. As the temperature drops below 2.18 K superfluidity sets in until, around 1 K, all of the liquid becomes superfluid. This effect arises because, at a sufficiently low temperature, all of the atoms fall into the same lowest possible quantum state and they all act together. If an attempt is made to change the state of motion of one of them, for example by introducing heat at a point, the states of all of them change in a similar way. It must be stressed here, however, that this superfluidity effect is quite different in nature and origin from superconductivity as discussed in section 6.4. Effects such as this in solid materials (e.g. in rubidium, lithium and sodium) have also been observed recently (1995) at amazingly low temperatures of the order of 10^{-7} K. At such temperatures the de Broglie wavelength (see section 5.3) of each atom is comparable to the spacing between the atoms and we have a *macroscopic* quantum system—the whole system, as well as its components, is a quantum entity. The importance of such systems for future technology is unpredictable at this stage.

6.7 Moving Forward

In this chapter we have seen how quantum ideas enable us to understand some important features of matter—the nature of the forces which hold matter together, its electric and magnetic properties and its strange behaviour when at very low temperatures. The next step is to delve into the nature of the atomic nucleus itself and to introduce the many basic elementary particles, of which the electron and photon are just two examples, which are the fundamental constituents of the physical world. However, we shall find that, in discussing these matters, we have to deal with situations where the particles have very high energies and speeds. Under these circumstances the mechanical theories we have used so far, both classical and quantum, cease to be accurate and relativistic ideas due to Einstein have to be introduced. This is our concern in the next chapter.

CHAPTER 7

EINSTEIN'S RELATIVITY THEORY

Space, Time, Mass and Energy

7.1 What is Relativity?

On reflection it will be recognized that the physical ideas and phenomena discussed in the previous chapters have essentially been concerned with the behaviour of moving pieces of matter—electrons, atoms, molecules—and the way these movements account for the properties of bulk matter. Both classical mechanics and quantum mechanics have been used to specify and understand this behaviour and in all situations the speeds of the various pieces of matter have been relatively small compared with the one high speed we have encountered, namely the speed of light ($c = 2.998 \times 10^8 \, \text{m s}^{-1}$—see section 4.4). At this juncture it should be recognized that we can only discuss speed in a *relative* sense. When we speak of travelling at $100 \, \text{km h}^{-1}$ along a motorway, this speed is measured relative to the motorway. However, the motorway itself is moving since it is fixed to the surface of the earth which is rotating, whilst the earth itself is travelling at high speed in orbit about the sun. In other words we discuss the car's motion in terms of a notional two-dimensional (left–right and forward–backward) *frame of reference* (see figure 7.1) and, in the above example, this frame is, as it were, fixed to the motorway.

Not all frames of reference are appropriate. For example, in discussing the general motion of everyday objects around us it is natural to use a three-dimensional (left–right, forward–backwards, up–down) frame notionally fixed to the surface, of the earth. However, in studying the motion of planets about the sun, because of the earth's own orbital motion, the motion of other planets would appear to be highly complicated if referred to a

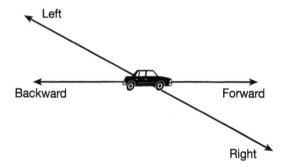

Figure 7.1: *A two-dimensional frame of reference.*

frame fixed to the earth. Here it would be much more appropriate to use a frame fixed to the sun and distant stars. With reference to such a frame all planetary motion would be simple orbits and be readily accountable for in terms of Newton's Laws of Motion (see section 2.3) and the force of gravity. In order to understand physical processes it is important to use what are known as *inertial frames of reference*. These are frames of reference such that a free particle subject to no force always moves in a straight line with uniform speed (which can, of course, be zero)—it obeys Newton's First Law of Motion (see section 2.1). A frame based on distant stars is a very good approximation to such a frame and, for most practical purposes, a frame fixed to the earth is satisfactory for studying earth-bound processes; the rotation of the earth and its movement about the sun only lead to very small corrections.

It is our general experience in everyday life that physical behaviour is the same in two frames of reference which are travelling at a uniform speed relative to one another. Physical processes in a train travelling at a steady speed are exactly the same as when the train is stationary. Of course absolutely steady travel is not possible and there are lurches and vibrations which disturb these processes. However, in a perfect train with a perfect track you could in principle clearly play a satisfactory game of billiards. This would not be so in a train that was accelerating or decelerating; *uniform* speed is essential. In other words it then might be conjectured that physical laws, both classical and quantum, are the same in *all* inertial frames of reference which are moving at a uniform speed with respect to each other.

There is one problem, however, and that concerns the speed of light. That it has a unique value, given above, is part and parcel of the theory of electromagnetism formulated by Maxwell (see section 4.4). This means that its speed should be the same with respect to all inertial frames of reference. However, our experience is that if, for example, you are travelling in a car at a uniform speed and shoot a bullet forward then, relative to the ground the bullet's speed will have been enhanced by the speed of the car (figure 7.2(a)). Similarly, when the car headlights are switched on, the speed of the light emitted relative to the ground should also be enhanced (figure 7.2(b)). That is, the speed of light would appear to be higher in the inertial frame fixed to the ground than in the frame fixed to the car in contradiction to the conjecture of the last paragraph. It was faced with this contradiction that Einstein in 1905 formulated his *special theory of relativity* which removed it. This theory is essentially embodied in two postulates both of which have already been foreshadowed.

1. *All the laws of physics are exactly the same in all inertial frames of reference.*

2. *The speed of light in empty space is the same in all inertial frames of reference.*

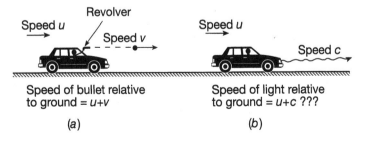

Figure 7.2: *(a) A bullet shot from a car; (b) light emitted from a car.*

The implications of this theory are profound and far reaching. They will be discussed in the next few sections.

7.2 Simultaneity

Two events are said to be simultaneous when they occur at *exactly* the same instant of time. This is straightforward to establish when both events take place at the same position in space, for example two trains arriving at the same time at a railway station. However, even in this example, the trains are not, fortunately (!), arriving at exactly the same position and to be sure that their arrivals coincided exactly the observer would need to station him/herself at the centre point between the two sets of buffers. If the trains touched the buffers at exactly the same instant the observer would *see* this to be so since, being exactly central, the light would take the same time to reach him/her from each set of buffers. The question then arises as to whether an observer in another inertial frame of reference would also agree that the trains had arrived together. To explore this let us leave trains and consider an idealized 'thought experiment'.

A thought experiment is a simple representation of a physical process unencumbered by unimportant peripheral details such as trains and buffers. Consider, in this instance, an observer standing at the centre of a room sending two simultaneous flashes of light to

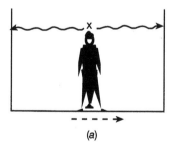
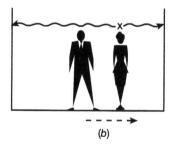

(a)   (b)

Figure 7.3: *(a) Two observers together at the instant two flashes of light are emitted; (b) the position of the observers at a later time.*

opposite walls of the room as shown in figure 7.3(a). Since he is at the centre of the room the flashes will reach the walls, according to this observer, at the same instant. But suppose at the instant the flashes were emitted that a second observer was passing the first observer travelling at a uniform speed, i.e. in another inertial frame of reference. The second postulate of relativity requires the speed of light to be exactly the same in her frame as for him, but, since she is in motion, the right-hand wall in her frame of reference is moving towards her, whilst the left-hand wall is moving away (see figure 7.3(b)). This means that in her frame the light has less distance to travel towards the right-hand wall and more towards the left-hand wall. It therefore arrives at the right-hand wall *before* the left hand wall and the two arrivals are no longer simultaneous. So we have a situation in which one observer says that two events happen simultaneously whilst the other says they do not!

Incidentally, this does not mean that it is possible for the second observer, having seen an event before it is due to be seen by the first, to predict the future for him since, by the time the information was communicated to him by her (most quickly by a light signal), the event would already have been observed. This means that for each observer *cause* still precedes *effect*. Neverthess, it has to be concluded that in relativity theory two events that are simultaneous for one observer are not simultaneous for another observer in a different inertial frame of reference.

7.3 Time Dilation

The foregoing discussion implies that there is nothing absolute about time. For any particular observer time appears to flow uniformly in relation to his/her frame of reference. However the time sequence of events for observers in different frames is, as we have seen, not always the same. This has a profound bearing on the measurement of time in different inertial frames. To understand this in more detail let us consider another thought experiment. Suppose the first observer measures the time taken for a light signal to be sent from him and reflected back in a mirror M fixed in his frame of reference (frame A) as shown in figure 7.4(a). The time which will have elapsed when the light has travelled to the mirror

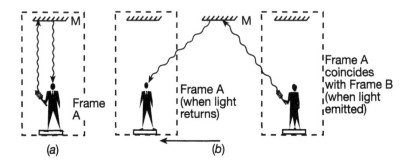

Figure 7.4: *The journey of a light signal (a) with reference to the first frame, and (b) with reference to the second frame.*

and back to him is simply the time taken to travel twice the distance separating him and the mirror. Now consider a second observer travelling at a uniform speed (moving to the right) relative to him. She could carry out the same measurement with an equivalent system set up in her frame of reference (frame B) and, using identical clocks, would obtain the same value for the time taken for the light to travel to and from a mirror set at the same distance from her. However, suppose she measures the time taken for the journey of the light in the frame A using the clocks in *her* frame of reference (B). From her point of view the mirror M in the first frame is moving with uniform speed to the left and so if she coincides in position with the first observer at the instant the light starts on its journey she will see it follow a longer path as shown in figure 7.4(b). This means that the time for the complete journey is longer when measured in relation to frame B. If the time interval for this journey were used as the 'tick–tock' of a clock (equivalent, for example, to the forward–backward motion of a pendulum), then the second observer would regard time as flowing more slowly in frame A than in her own (frame B) since it takes longer for a 'tick–tock' to be completed. We have here the phenomenon of *time dilation*. Of course, the situation is completely symmetrical since motion is relative and the first observer, carrying out the equivalent measurment, would equally regard time as flowing more slowly in frame B. At first sight this raises a problem which is famously summarized in what has been called the *twins paradox* as follows.

Although the foregoing discussion has been carried through by considering time in relation to processes involving light signals, the conclusions about time dilation hold generally and not simply for these specific processes. In other words, all types of clock— mechanical or electrical, biological processes etc are subject to time dilation effects. These effects, however, are very small unless relative speeds near to the speed of light are involved; they have no effect on our everyday experiences. They have been demonstrated most effectively by measuring the time taken for radioactive particles (see section 8.6) to decay when at rest and when in high-speed motion in the laboratory. It is found that their lifetimes in motion are increased precisely as predicted by relativity theory. Turning now to the twins paradox, this simply considers the ageing implications for a pair of twins, one of which remains on earth whilst the other makes a high-speed round trip into space. Relative to the twin on earth, time will have moved more slowly (will have been dilated) for the twin in the spacecraft and the travelling twin should have aged less than the earthbound twin. However, in relation to the spacecraft's frame of reference, it is the earth that has moved away and then returned so the travelling twin would expect the earthbound twin to have aged less. The resolution of this paradox is to recognize that the situation is not completely symmetrical. The frame of reference based on the earth is, as mentioned in section 7.1, essentially an *inertial frame* whilst, with reference to this frame, the spacecraft experiences accelerations during its launching, turn around and landing. The earthbound twin is always in an inertial frame whilst the travelling twin is not. Detailed considerations* then show that the first conclusion—that the travelling twin has aged less—is correct.

7.4 Length Contraction

An important implication of time dilation is that length in one inertial frame of reference (A) measured by an observer in another inertial frame of reference (B) moving relative to it is *less* than the length which would be measured by an observer actually in frame A. This can be seen most simply by considering a radioactive

---

* See, for example, Davies P 1995 *About Time* (New York, Simon and Schuster).

particle travelling in a fixed frame of reference, for example a laboratory. An observer in the laboratory will see it travel a certain distance before it decays. This distance is simply its speed multiplied by its lifetime *as measured in the laboratory* which, remember, is longer than its lifetime if it were at rest. However, an observer travelling with the particle sees the same span of laboratory pass at the speed of the particle but, since the particle is at rest relative to him/her, its lifetime is shorter than that measured in the laboratory. In turn, the distance this observer sees covered in the laboratory is also shorter. Thus, length in the laboratory as measured by the moving observer is less than length measured by the fixed observer—*length contraction* has taken place because of the motion. Distance, like time, is not absolute—it depends on the frame of reference from which it is measured, but, again, as with time this is a relativistic effect and plays no part in the nature of mechanics and motion unless the entitities concerned are travelling with speeds comparable to the speed of light.

Clearly, however, if these effects are taken into account, then the associated mathematical theory will differ from that describing non-relativistic phenomena. Nevertheless, it should be expected that the mathematics describing motion should revert back to classical form when dealing with low-speed phenomena. To illustrate this consider the way in which speeds were added together in figure 7.1. Here a problem arose when adding the speed of the car to the speed of light since it gave a value greater than the speed of light in the frame of reference fixed to the ground which is forbidden by relativity theory. It turns out that the relativistic formula for the addition of two speeds u and v is rather different from the classical formula. It is

$$\text{total speed} = (u + v)/(1 + uv/c^2).$$

If u and v are much smaller than c (the speed of light) then the second term in the denominator can be neglected and we revert back to the simple expression $u + v$ for the total speed. At the other extreme, if v, for example, is equal to c, then it requires no great algebraic skill* to see that the total speed is simply c as

---

* Total speed = $(u + c)/(1 + uc/c^2) = (u + c)/(1 + u/c) = c(u + c)/(u + c) = c.$

required. Whatever speed is added to c, the resultant speed is still c!

7.5 Mass and Energy

The two postulates of relativity (section 7.1) and the discussion of some of their implications leads on to a remarkable relationship between energy and mass. In sections 2.2–2.4 we saw that when a force is applied to an object its motion in the direction of the force accelerates and it gains an amount of kinetic energy equal to the work done by the force; energy is conserved. Continuing to exert the force, the speed of the object steadily increases and classically there is no limit to the speed that can be attained.

However, this is not the case with relativity theory; the maximum attainable speed is c, the speed of light. As the speed approaches c it becomes harder to accelerate the object. Yet, if energy is conserved, the work being done by the force must in some way be increasing the energy of the object. This energy is dependent on only two quantities—the object's mass and its speed. If the latter cannot readily increase in response to the force it follows that the former must somehow be increasing and the nearer the speed becomes to c the larger will be this effect. We thus come to the remarkable conclusion that the mass of an object increases as its speed increases, becoming very large indeed (heading for infinity) as this speed approaches the speed of light. Clearly the lowest value this mass can have is when the object is at rest and this value is referred to as the object's *rest mass* (denoted by m_0). The increase in the mass of an object due to its motion is proportional to its rest mass and it therefore follows that the only circumstance in which there is no increase in mass is if the rest mass is itself zero. This is the situation with a photon and that is why, to state the tautologically obvious, it does travel with the speed of light. We shall see in Chapters 8 and 9 that there are other elementary particles that are also believed to have zero rest mass and which also, therefore, always travel at the speed of light.

Further, since energy (provided by the work done by the force) is being transformed into mass, it follows that there must be an intimate relationship between these two physical quantities. It turns out that this relationship is extremely simple and is embodied in probably the best known of all physical formulae, namely

$$\text{energy} = (\text{mass}) \times (\text{speed of light})^2$$

or, more succinctly,

$$E = mc^2$$

where E represents energy and m mass. In principle, then, energy can be converted into mass and mass into energy. This being so, means that it is no longer sensible in the relativistic context to talk about the 'conservation of energy' since clearly energy and mass are equivalent. Rather we have to employ a more embracing law—*the conservation of mass–energy*. All this being said does not imply that our considerations in earlier chapters about motion, energy and the behaviour of matter in the everyday world are invalidated. To be sure they are not *exactly* right but, since the speeds involved are exceedingly small compared with the speed of light, any corrections to our conclusions are absolutely negligible.

Reverting back to the moving object we have been discussing, its momentum (p) is simply given by the classical expression *mass* $\times$ *velocity* except that, instead of using what we have now called rest mass (m_0) for the mass of the object, the relativistic mass (m) must be used. In these terms, a very simple relationship for the total energy (E) of an object, including its rest-mass energy, can then be obtained. It is

$$E^2 = m_0^2 c^4 + p^2 c^2.$$

Thus, for a particle at rest $(p = 0)$ the energy contained within its mass is

$$E^2 = m_0^2 c^4 \quad \text{or} \quad E = m_0 c^2.$$

At the other extreme, for an object of zero rest mass $(m_0 = 0)$, such as a photon, the relation between energy and momentum is simply

$$E^2 = p^2c^2 \quad \text{or} \quad E = pc.$$

Since for a photon $E = h\nu$ where ν is its frequency and h is Planck's constant (see section 5.2), we have

$$h\nu = pc.$$

Then, using the relation between frequency, wavelength and wave speed (see section 2.6) the expression relating wavelength, momentum and h used in section 5.3 is obtained, namely

$$\text{wavelength} = c/\nu = h/p.$$

Since the speed of light is so large, the mass–energy relationship does imply that if a small amount of mass could be converted into energy the amount released would be extremely large indeed. To put this into proportion, 1 g of matter completely converted into energy could run a one-bar electric fire 24 h a day for a year in around 3000 homes! Unfortunately, as we shall see in Chapter 8, it is not possible to completely annihilate a piece of matter and convert all of its mass into energy. This can only be done with relatively small quantities. Even so, when this is arranged the energy release can still be very large—witness nuclear explosions brought about by *fission* or *fusion* (see section 8.5).

7.6 Relativistic Quantum Mechanics

We have seen that the postulates of relativity theory have required modifications in the way in which we deal with the classical treatment of motion, energy and momentum. In particular the three space and one time dimensions turn out to be intimately related. No longer can time be simply regarded as an 'ever-flowing stream' unrelated to spatial considerations. Indeed we are in a situation where events must be dealt with in a more general four-dimensional space including time. For two inertial frames moving with uniform speed relative to each other we have seen that the four coordinates specifying spatial position and time in one frame are related to the coordinates in the other frame in a way that is more complex than obtained by simple classical considerations.

It is the requirement that all physical laws should be the same in all inertial frames of reference that has led to the modifications just referred to. The resulting relativistic laws of motion are then said to be *invariant* under a transformation which changes the space–time coordinates according to the above relations. This transformation of coordinates is referred to as the Lorentz transformation, named after a Dutch physicist who first formulated the details of the transformation. In turn, we should also expect that this invariance requirement also has implications for the theory of quantum mechanics. Here it turns out that the Schrödinger equation from which wavefunctions are derived (see section 5.4) is not Lorentz invariant and, therefore, if relativistic effects are to be dealt with, it has to be modified.

The problem is that the Schrödinger equation—a complicated differential equation—is based on the classical relationship between energy and momentum, namely (see section 2.4)

$$E = p^2/2m_0$$

rather than the relativistic expression given in the last section. Naturally, with the advent of relativity theory, relativistic quantum mechanical wave equations were developed. Two even more complicated equations emerged. One was based directly on the relativistic expression relating energy and momentum given in the last section and is known as the *Klein–Gordon Equation*. The other was based on the *square root* of that equation, namely

$$E = \sqrt{(m_0^2 c^4 + p^2 c^2)}$$

and is known as the *Dirac Equation*. This latter equation, which was brilliantly derived and developed by the British physicist and Nobel Laureate, Paul Dirac, led to remarkable outcomes.

First, it is found that the equation only makes sense if it describes the behaviour of particles having an intrinsic spin of $\frac{1}{2}$. This means that it is the natural equation to use when dealing with electrons which have spin $\frac{1}{2}$ (see section 5.5): the spin of the electron is, as it were, built into the mathematics and does not have to be tacked on. Second, the magnetic properties of the electron (see section

5.5) predicted by the equation agree almost exactly with those observed. Third, the equation also has solutions for negative energies. This comes about because when the square root of an expression is taken, as above, the result can be negative as well as positive ($2 \times 2 = 4$ and, also, $(-2) \times (-2) = 4$). This raised a terrible problem because any normal electron with a positive energy would drop into one of these negative-energy states and release energy in the form of electromagnetic radiation (a photon) in just the same way that electrons emit radiation in an atom by jumping from high- to low-energy states (see section 5.6). Since there is an infinite number of negative-energy states with energies ranging from $-m_0 c^2$ (when p was zero) downwards (as p increases) it follows that no electron could exist with positive energy; it would always fall into this bottomless pit of negative-energy states.

Dirac's approach to solving this problem was to propose that *every* negative-energy state is occupied by an electron. In other words, so-called empty space—referred to in physics as the vacuum—is in fact filled with an infinite 'sea' of unobservable negative-energy electrons. The point of this proposal was that a positive-energy electron would then be unable to drop into a negative-energy state because the Pauli exclusion principle (see section 5.5) forbids two electrons occupying the same state and the negative-energy states are all occupied. The situation is illustrated in figure 7.5(a) where a single electron sits above the infinitely deep sea of occupied negative-energy electron states. Here it will be noted that there is a gap of magnitude $2m_0 c^2$ which cannot be occupied by a free electron since it would imply that its rest mass was less than m_0. (With $p = 0$, $E = + m_0 c^2$ or $- m_0 c^2$.)

This proposal then had a profound consequence. Considering only the 'vacuum', suppose a photon is absorbed by a negative-energy electron and it is given sufficient energy (greater than $2m_0 c^2$) to raise it into a positive-energy state as shown in figure 7.5(b). The resultant 'hole' in the negative-energy sea will behave as a positively charged electron as we saw when discussing holes in the context of semiconductors in section 6.3. Thus, whereas before the photon is absorbed by the negative-energy electron there is just the vacuum, after the excitation there is a pair of additional particles—a negatively charged electron and a positively charged

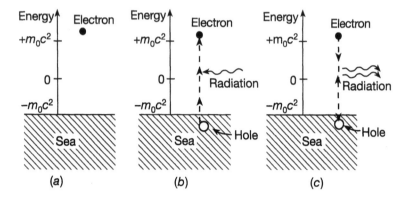

Figure 7.5: *The sea of negative-energy electrons. (a) With a single positive-energy electron; (b) a negative-energy electron excited into a positive-energy state, leaving a hole; (c) the annihilation of a positive-energy electron by a hole.*

'electron'. The process is known as *pair production*. This latter particle is now called a *positron* (denoted by e⁺) and is one example of what are referred to as *antiparticles*; they will be discussed more fully in Chapter 9. It has exactly the same mass as an electron and has the same spin ($\frac{1}{2}$) but a positive charge equal in magnitude to that of the electron. The prediction that they should exist was a tremendous triumph for Dirac when his conjecture was confirmed experimentally by Anderson in 1932 who first detected positrons in cosmic rays (particles impinging on the earth from outer space) by analysing the tracks made by them in photographic plates. Since then, pair creation and the properties of positrons have been studied in great detail in the laboratory. There is also an inverse process that can take place shown in figure 7.5(c), in which an electron falls into a hole, the released energy being carried away as electromagnetic radiation (in this case, to satisfy conservation laws, *two* photons are emitted). Both electron and positron disappear and this process, for obvious reasons, is known as *pair annihilation*.

It should also be remarked that the production and annihilation of particle–antiparticle pairs is continually happening in empty space. In section 5.6 the uncertainty relation connecting energy and time was mentioned—for short time periods the energy is not precisely known. Energy, as it were, can be 'borrowed' for a while but must be handed back after the time period (10^{-20} s or less) has elapsed. Thus energy to create a particle–antiparticle pair is available for that short period and empty space is full of these transitory pairs.

The concept of an infinite sea of negative-energy particles in the vacuum is clearly difficult to assimilate and, since Dirac's original formulation, the theory of particles and antiparticles and their creation and annihilation has been considerably developed so that no longer does understanding rest on such a strange physical idea. The Dirac equation is still used to descibe particles of spin $\frac{1}{2}$ but in a somewhat different way and developments in what is called *quantum field theory* now enable us to deal in a much neater and logical way with the creation and annihilation of particles and antiparticles without having to retain the concept of a negative-energy sea of particles. Similarly, the Klein–Gordon Equation is used in this context to describe particles having, for example, spin 0.

7.7 General Relativity

The special theory of relativity and its use in the context of classical and quantum mechanics was a revolutionary leap forward in the understanding of a wide variety of physical processes, particularly those involving high energies and speeds. However, it did not fully encompass gravitational phenomena. The problem is that the Newtonian theory of the gravitational force between two bodies (see section 2.2) presupposes that the force between them is transmitted instantaneously—if the mass of one body is increased the other should *immediately* experience an increase in the attractive force. This implies that the force is being transmitted faster than the speed of light, which is not acceptable. Further, special relativity is only of use in the context of inertial frames of reference: relative motion when one of the frames is accelerating is not taken on board.

It was to deal with these problems that in 1915 Einstein formulated what is referred to as the *general theory of relativity*.

It was encompassed in two main postulates. The first was as follows:

1. *All the laws of physics can be expressed in such a way that they have the same form in all frames of reference whatever their motion.*

Accelerated relative motion of frames of reference is now incorporated, but to achieve this means that the laws have to be expressed in a much more complicated mathematical form. The search still goes on for ways of doing this for all the physical interactions that are known, not least the electromagnetic interaction.

The second postulate related to the apparent local equivalence of the gravitational force and acceleration. Everyone has experienced the sensation of increased weight in a lift as it accelerates from rest and goes upwards. The force exerted on the floor of the lift by our bodies increases whilst the lift is accelerating as though the force of gravity due to the earth had increased. The second postulate of general relativity effectively states that in a sufficiently small closed accelerating container, such as a lift, from which there is no opportunity to make an external observation to confirm that it is accelerating, there is no way within the container of determining whether the apparent increase in weight is due to acceleration or to an increase in the force of gravity. The two possible causes of the increase are equivalent. Indeed the second postulate is referred to as *the principle of equivalence* and can be stated as follows.

2. *There is no way for an observer in a sufficiently small closed laboratory to distinguish between the effects of a gravitational force and those produced by accelerated motion.*

It is this equivalence that leads to the identity between gravitational and inertial mass mentioned in section 2.2. Clearly, within the general theory of relativity, the gravitational force is treated on a different footing from other types of force. In fact the complex mathematics of general relativity leads to the inescapable conclusion that in the vicinity of matter or energy the nature of space–time is modified; it becomes curved or warped. To understand this in terms of our everyday experience is difficult if not impossible, but analogies can help. Imagine a two-dimensional being whose only experience of space is motion forwards–backwards or left–right on a flat surface (see figure 7.6(a)). In this case the shortest distance of travel between two points is a straight line. Introduction of a gravitating body will, according to general relativity, distort the surface as in figure 7.6(b) and the shortest distance of travel on this curved surface is now no longer a straight line. Similarly elementary (Euclidian) geometry, for example Pythagoras' theorem, no longer holds on this curved surface—a triangle is no longer constituted from straight lines. With that analogy in mind the motion of a piece of matter in a gravitational field (e.g. planetary motion) is now interpreted as due to an intrinsic curvature of space and time; a planet in orbit simply moves along the shortest path in curved space–time (the equivalent of a straight line in 'flat' space–time—known as a *geodesic*) and this appears as its orbit in our everyday three-dimensional space.

It turns out that planetary orbits calculated using general relativity are virtually the same as those calculated using simple classical mechanics and Newton's form of the gravitational force. However there is one measurable difference for the orbit of Mercury, the planet nearest the sun and, therefore, experiencing the strongest gravitational field. Due to the influence of other planets, its perihelion (the point of its orbit nearest the sun) precesses about

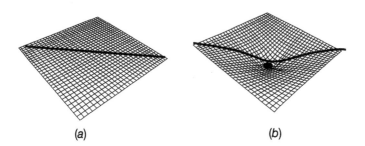

(a)   (b)

Figure 7.6: *Motion on a two-dimensional surface (a) when flat and (b) when curved.*

the sun as shown in figure 7.7(a). The magnitude of this precession is a little less than 2° per century, but the measured value is slightly different from the calculated value by somewhat less than one-hundredth of a degree per century and it was a tremendous triumph for the general theory of relativity when it was shown that it could account exactly for the discrepancy.

The theory made two other testable predictions. One was that due to the curvature of space–time the motion of a light ray passing through a gravitational field should deviate from the usual straight line motion. This is a small effect and can only be tested when the light passes near a very large gravitating body such as the sun. It was checked in 1919 by studying the apparent position of several stars whose light was passing near to the sun (see figure 7.7(b)). Clearly this could only be done during a solar eclipse otherwise the light from the sun overwhelms that from the star. It was found that the apparent position of such a star did indeed differ from its observed position at other times when its light did not pass near the sun and that the measured deviation agreed well with the theoretical prediction.

Another prediction was that, because of the curvature of space–time, the rate of a clock in a high gravitational field is slightly slower than when in smaller gravitational field. This is an

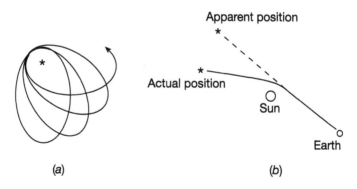

Figure 7.7: *(a) The precession of the perihelion of Mercury. (b) The curved path of a light ray near the sun.*

additional form of time dilation to that due to motion discussed in section 7.3 in the context of special relativity. Clocks should run more slowly on the surface of the earth than up in the atmosphere or in outer space. The effect is very small—of the order 1 part in 10^9. Even so it was confirmed for atomic clocks (clocks for which the unit of time is the time taken for one vibration of the electromagnetic radiation emitted by an atom—usually caesium) carried on 'round the world' journeys during 1971–2 by a commercial airliner at high altitude (around 10,000 m). To check the theory it was also necessary to take into account the special relativity time dilation resulting from the airliner's motion. Other checks have been made more recently using measurements in a spacecraft and also by studying the frequency of light from atoms at the surface of the sun, where there is a strong gravitational field, compared with that of light emitted by identical atoms on the earth where the gravitational field is much weaker.

A final prediction was that moving objects, especially heavy ones, would cause ripples in space–time analogous to a moving object making ripples in a pond. These ripples are referred to as *gravitational waves* and they are rather like electromagnetic waves produced by moving electric charges. This effect would lead to a moving body losing energy by gravitational radiation. However the effect is very small indeed and, so far, no gravitational waves have been detected. However, the resultant energy loss has been measured in rotating pairs of neutron stars (see section 10.5). Here it should also be noted that the quantized version of this gravitational radiation, possibly equivalent to the photons associated with electromagnetic radiation, and known as *gravitons*, enter theories incorporating *all* the elementary particle interactions (see section 9.4), but, unlike photons which have spin 1, gravitons have spin 2.

One interesting feature of the general theory of relativity is that, as first formulated by Einstein, it predicts that the universe is in a state of continual expansion. As will be discussed in Chapter 10 this does, indeed, seem to be the case. However in Einstein's time it was believed that the universe was static. For that reason he introduced a constant, known as the *cosmological constant*, into the theory; this effectively cancelled the expansion. Since the general theory does allow the presence of such a constant and since the universe is expanding, its precise value is now subject to much discussion. We shall return to this issue in section 10.3.

7.8 Moving Forward

This chapter has been, as it were, an interlude in our discussion of the tangible physical world. It has introduced ideas and concepts about space and time that challenge our everyday experience and our powers of imagination. The special and general theories of relativity are powerful and very necessary when considering phenomena at high energy and, remembering $E = mc^2$, large accumulations of matter and their associated gravitational fields. However the relation between mass and energy plays an important role at the microscopic level as we shall see when considering the behaviour of atomic nuclei and elementary particles. The former is the focus of our discussion in the next chapter.

CHAPTER 8

THE ATOMIC NUCLEUS

Protons, Neutrons and their Interactions

8.1 Nuclear Constituents

The work of Rutherford, Geiger and Marsden described in section 5.1 established the existence of atomic nuclei and that they are very small (of the order 10^{-14} m for gold atoms). Nuclei have positive electric charge to hold the negatively charged atomic electrons in their 'orbits' and, since atoms are electrically neutral, this nuclear charge must exactly balance the total negative charge carried by the surrounding electrons. If the number of electrons in an atom is denoted by Z (the *atomic number*) and the electron charge by $-e$ it follows that the nuclear charge must be $+Ze$. Most of the mass of an atom is in its nucleus, the mass of the lightest (hydrogen) being around 1840 times the mass of an electron. The masses of heavier nuclei turn out to be roughly equal to integer multiples of the mass of the hydrogen nucleus. This integer is known as the *mass number* and is denoted by A. It takes the value 1 (for hydrogen), 4 for helium and so on through to around 260 for the heaviest elements. A is usually of the order two or more times the value of Z.

These observations suggested first of all that atomic nuclei consist of two types of component—hydrogen nuclei (now called *protons*—denoted by p) and *electrons* (denoted by e$^-$). To have the mass roughly right there would need to be A protons and since the charge of each proton is $+e$ there would need to be sufficient electrons (charge $-e$) in the nucleus to reduce the total proton charge ($+Ae$) to $+Ze$. However, simple as this idea is, it cannot work since, confining an electron to a nucleus of size 10^{-14} m means that the uncertainty in its position is very small. In turn, the quantum mechanical uncertainty relation (see section 5.3) implies that the uncertainty in its momentum (and therefore its kinetic energy) is very large. It turns out that this kinetic energy is so large that the electrical attraction between the positively charged protons and the negatively charged electrons is far too weak to confine the electrons in the nucleus; the quantum mechanical agitation due to their confinement means that they would simply break out of the nucleus.

The solution to this problem finally emerged in 1932 when an experiment by Chadwick established the existence of a new electrically neutral particle having virtually the same mass as a proton. This particle is known as a *neutron* (denoted by n) and an atomic nucleus can then be regarded as consituted from Z protons (to get the charge $+Ze$ right) and $N = A - Z$ neutrons (to get the mass right; $Z + N = A$). The idea that such particles might exist had been mooted by Rutherford some years earlier. Chadwick established their existence in an experiment in which alpha-particles (see section 5.1) from a radioactive source bombarded a piece of beryllium. It was already known that this led to the emission of some form of very penetrating electrically neutral radiation. Chadwick studied this radiation by allowing it to fall on paraffin wax (a hydrogen rich substance) from which it knocked out hydrogen nuclei (i.e. protons) as shown in figure 8.1. These protons were then detected in what is known as an *ionization chamber*. The latter is a chamber containing gas and two electrodes. When a proton enters the chamber it knocks electrons out of the gas atoms leaving positively charged ions (see section 6.1). An electric potential difference between the electrodes then results in a burst of electric current in an external circuit due to the presence of these charged ions, so indicating the arrival of a proton. By studying this process Chadwick established that the penetrating radiation that

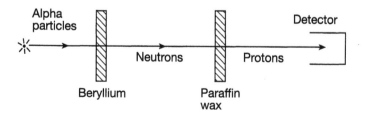

Figure 8.1: *The Chadwick neutron experiment.*

unleashed the protons was indeed composed of particles (neutrons) having a mass very close to that of protons.

If the nucleus consists of neutrons and protons it follows that, for a particular chemical element—defined by the number of electrons in the atom—the mass of the nucleus can take various values. For an atom having Z electrons the nucleus must contain Z protons but the number of neutrons could vary. In practice this variation is limited and there are only a vew possibilities which lead to a stable or fairly stable structure for the nucleus. The atoms containing these different possible nuclei have identical chemical properties since they have the same number of electrons. They are referred to as *isotopes* and those that are only 'fairly' stable are radioactive.

The foregoing description of atomic nuclei is straightforward and satisfying. The nucleus simply consists of an assembly of Z protons and N neutrons, the two types of particle being referred to collectively as *nucleons*. A particular nucleus X (say) is defined by the two numbers A and Z and denoted symbolically by ^A_ZX. For example $^{12}_{6}$C and $^{13}_{6}$C represent stable carbon isotopes, $^2_1$H represents what is known as the heavy hydrogen nucleus and so on. However this description has a profound implication. There must be a very strong attractive force holding the nucleons together in the nucleus, not least because the electrical repulsion between the component protons, which are very close together, has to be withstood. In section 8.3 we shall consider the nature and origin of this *nuclear force* which is much stronger than the electric force and which is one example, as we shall see in Chapter 9, of what is known as the *strong interaction*, but before discussing this force a few of the general properties of nuclei for which the force must account will be described.

8.2 General Properties of Nuclei

Nuclei are studied and their nature and properties revealed in essentially three ways. First, they can be bombarded by other particles to see how these particles are scattered or how the nucleus disintegrates. We speak of the latter process as a *nuclear reaction.* The nuclear behaviour in such processes is dependent on its structure and careful analysis of this behaviour can reveal a great deal of information about the structure. Second, many nuclei are radioactive and the nature of this radioactivity again depends on the structure of the nuclei involved. Third, the fine details of atomic spectra or the behaviour of nuclei in applied electromagnetic fields gives information about their electric and magnetic properties. In subsequent sections these different investigative processes will be discussed in more detail. In this section the focus of the discussion will be simply on the general properties of nuclei which emerge from such studies.

Shape and Size. Nuclei are frequently spherical but sometimes have shapes rather like a lemon or a squashed orange. Their radii vary from around 10^{-15} m for the lightest through to around 10^{-14} m for the heaviest and their volumes turn out to be simply proportional to the value of A, i.e. proportional to the number of particles they contain. They are, of course, very dense since virtually all the mass of an atom is concentrated in their very small volumes. This density is some 10^{14} times greater than that of ordinary matter—one cubic inch of nuclear matter would weigh around 4,000,000,000 tons! Finally, the edge of a nucleus is not sharp like a billiard ball but is somewhat 'fuzzy'.

Binding Energy. As has already been mentioned there must be a strong attractive force between nucleons which holds them tightly together in the nucleus. Given such a force means that if a nucleon is to be extracted or knocked out of a nucleus then work has to be done against this retaining force—energy has to be provided. Of course, the amount needed varies, roughly speaking, according to whether the nucleon being extracted is hovering near the nuclear surface or is buried deep in the interior. It obviously varies from nucleon to nucleon but will have an average value. This average value is simply the energy needed to completely knock a nucleus to pieces so that all the nucleons are well separated from each other, divided by the total number of nucleons (i.e. A). The energy needed to disintegrate the nucleus in this way is referred to as its *binding energy* (denoted by B) since it is the energy binding the nucleus together. In turn, it follows that if we brought all the component nucleons together from a large distance an amount of energy equal to B would be released. This means, because of the equivalence between mass and energy (see section 7.5) and since energy has been *released* in assembling the nucleus, that its mass is *less* than the mass of the component nucleons by B/c^2. This is why nuclear masses, as mentioned earlier, are only approximately equal to the total mass of their component nucleons. It is interesting to plot the binding energy per nucleon (B/A) as a function of A as shown in figure 8.2. The energy is measured in MeV (millions of electron volts). The electron volt is a unit of energy widely used in physics and which is defined in the Glossary (see also section 4.2). We shall see later on that the shape of this curve has implications for the nature of the nuclear force and also for the processes of fission and fusion.

Nuclear Energy Levels. As with the electrons in an atom so the motions of nucleons in a nucleus must satisfy the laws of quantum mechanics. In turn, only certain energies are allowed for the nucleus and each nucleus is characterized by a series of possible excitation energies it can have—its *energy levels*. The lowest of these is referred to as the *ground state* and the remainder as *excited states*. Each state, as well as having a definite energy, also has a definite angular momentum and that of the ground state is

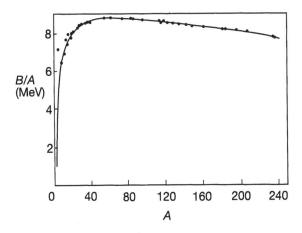

Figure 8.2: *Binding energy per nucleon (B/A) as a function of mass number A.*

known as the nuclear spin. It is found that those nuclei for which A is even always have states with angular momentum equal to an integer multiple of $\hbar$, and those with an odd value of A have angular momentum equal to $\frac{1}{2}, \frac{3}{2}, \frac{5}{2}, \ldots$, etc times $\hbar$. This latter observation has the implication that nucleons (protons and neutrons), like electrons, must have spin $\frac{1}{2}$. Being identical spin $\frac{1}{2}$ particles, protons obey the Pauli exclusion principle (see section 5.5) and so, similarly, do neutrons.

Electromagnetic Properties. It is to be expected that since nuclei contain charged protons which are moving about in the nucleus, like electrons in an atom, then the resultant electric currents lead to the nucleus behaving like a magnet. Such magnetic properties have been measured for the ground and some of the excited states of many nuclei. It is found that all nuclei having non-zero spin behave as magnets but the strength of this magnetism can only be understood if the protons and neutrons themselves, like electrons, also behave as miniscule magnets. The fact that nuclei with zero spin do not behave as magnets is easy to understand since, with no spin, the nucleus has no spatial direction associated with it whereas a magnet (think of a bar magnet) is highly directional. In addition, those nuclei which are non-spherical in shape produce a rather complicated electric field due to the associated non-spherical distribution of charge carried by the protons. These distorted charge distributions have been measured carefully and we shall see in section 8.4 that where the distortion is large it has an important effect on the nuclear energy levels.

Having briefly described some nuclear properties we now consider their implications for the nature of the nuclear force.

8.3 The Nuclear Force

A key feature of the nuclear force emerges by considering the implications of the way in which the binding energy per nucleon (B/A) varies with A as shown in figure 8.2. Imagine starting with one nucleon (say, a proton) and adding additional nucleons to build increasingly larger nuclei. From the curve it is clear that, to begin with, the binding energy per nucleon increases steadily. This can be easily understood because any one nucleon experiences the attraction due to the nuclear force of an increasing number of neighbours as they are added, so it becomes more tightly bound. However, the process does not continue and the B/A curve soon flattens off. Put this another way: for all but very light nuclei the average energy to remove a nucleon is much the same (around 8 MeV) whatever the size of the nucleus.

Contrast this with the energy needed to remove a piece of matter from differently sized gravitating bodies—much less energy is needed to launch a rocket from the moon than from the earth. This is because the rocket experiences the gravitational attraction of the *whole* body; the gravitational force, like the electric force (see section 4.1) dies away slowly with distance and can be experienced over very large distances—it is a *long-range* force. This cannot be so with the nuclear force. It must have a *short range* so that, in removing a nucleon from a nucleus, energy only has to be provided to overcome the force of attraction due to its immediate neighbours, the number of which is roughly the same for all nuclei except for the lightest. Any one nucleon does not experience the attractive force of more distant nucleons. The range of the nuclear force must, therefore, be less than the size of an average nucleus and can be estimated to be around 10^{-15} m. Here it should be remarked that it is the long-ranged nature of the repulsive electric force between protons, so that all the protons in a nucleus experience it, that inhibits the inclusion of too many protons in a nucleus. As nuclei become bigger this repulsion also reduces the binding energy B and leads to the gradual drop in the value of B/A as A, and therefore Z, moves to very high values (see figure 8.2).

Since nuclei increase in size the more nucleons they contain it also follows that, however powerful the nuclear force, nuclei do not collapse under its influence—nucleons keep their distance. The force must therefore have a very short-range repulsive element which does not allow the nucleons to approach very closely to each other. The nuclear force must therefore have a general form similar to that of the interatomic potential illustrated in figure 3.1 except that the strength and range are utterly different. The interatomic force has a range measured in atomic sizes (of the order of 10^{-10} m) as against the nuclear force range (10^{-15} m). The depth of the interatomic potential well (see figure 3.1) is measured in electron volts (eV) whereas the nuclear potential well must inevitably be of the order of the average binding energy per nucleon, namely measured in millions of electron volts (MeV). Very roughly speaking the nuclear force is a million times stronger and a million times shorter ranged than the interatomic force.

The fine details of the nuclear force have been exhaustively established by studying the way in which neutrons and protons behave when they are scattered one from the other like billiard balls. This involves accelerating protons by subjecting them to powerful electric fields in an *accelerator* or using neutrons from nuclear reactions such as that used by Chadwick in the experiment described in section 8.1. Information has also been derived from studies of the lightest composite nucleus (the heavy hydrogen nucleus—known as the *deuteron*) consisting simply of a proton and a neutron bound together. Here, for example, it emerges that the spins of the proton and neutron (each has spin $\frac{1}{2}$) in this nucleus are parallel so that the total spin is 1; when their spins are antiparallel the force is not strong enough to hold them together. This is a simple example of the extreme complexity of the nuclear force, which turns out to be highly dependent on the directions in which nucleon spins are pointing—not only in relation to each other but also in relation to their directions of motion.

One other important feature emerges, namely that the nuclear force between two neutrons, two protons or a neutron and a proton is exactly the same when the two particles are in the same state of motion (quantum state). Of course, since protons have an electric charge, its effects have to be subtracted out first to arrive at this conclusion. The implication is that neutrons and protons are intimately related and can be regarded as two different manifestations or states of the particle which, as has been mentioned, we call the *nucleon*. Although they have very different electromagnetic properties—charge and magnetic behaviour—they behave identically as far as the *strong* nuclear interaction is concerned. This identity in behaviour is reflected in the energy levels of what are called *mirror* nuclei. These are nuclear pairs such as the oxygen nucleus (eight protons, nine neutrons) and the fluorine nucleus (nine protons, eight neutrons). It is found that their energy levels are essentially the same, any differences being attributable to the electric charges of the protons.

Finally, we turn to the origin of the nuclear force. The electric force between, for example, two electrons can be understood in classical terms as due to the electromagnetic field between the electrons. This field, as we saw in section 4.4, is propagated with the speed of light and, in quantum terms, the transmission of the field is by means of photons. Photons, as we have seen, behave like particles carrying momentum and energy but having zero rest mass. Thus a picture emerges in which the electric force between two electrons is understood in terms of the exchange of photons (symbol γ) between them as illustrated in figure 8.3(a) where the photon is represented by a squiggly line. The photon is created by one electron and annihilated by the other. In some ways the interaction between them is rather like that between two tennis players who influence each other's movements—attraction and repulsion—by how the ball is exchanged between them. The formal quantum mechanical theory describing such an interaction between charged particles is known as *quantum electrodynamics*. It is called a *gauge* theory and the photons, which have spin 1, are also referred to under the collective name *gauge bosons*.

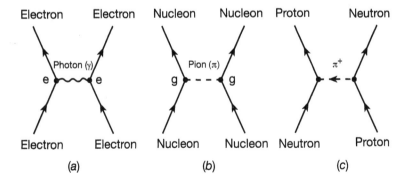

Figure 8.3: *(a) Photon exchange between electrons; (b) pion exchange between nucleons; (c) charged pion exchange between nucleons.*

In 1935 a Japanese physicist—Hideki Yukawa—suggested that the short-range nuclear force could be understood in a similar way but with the exchange of a finite-mass particle between nucleons instead of a photon. Just as photons are created and absorbed by charged particles, so this newly proposed particle is created and absorbed by nucleons. Of course, for such a particle to be created by a nucleon, energy $E = m_0c^2$ (see section 7.5) would be needed, where m_0 is the rest mass of the particle. Where does such a large amount of energy come from? Here it must be remembered that in quantum mechanics there is always a short period during which energy does not need to be conserved—there is an uncertainty about its value. This uncertainty is embodied in the uncertainty relation discussed in section 5.6. If we trade on this energy uncertainty and equate it to m_0c^2, then, from the uncertainty relation, it follows that the short time for which this energy is available is of the order $h/(m_0c^2)$ where h is Planck's quantum constant. The maximum speed with which the particle can travel is the speed of light, c, and hence the furthest distance it can travel is given by the product (speed $\times$ time), namely $c \times h/(m_0c^2) = h/(m_0c)$. If this distance is to be the range of the nuclear force (around 10^{-15} m) then it requires m_0 to be a few hundred times the mass of an electron (remember nucleons are around 1840 times the mass of an electron).

No such particle was known when this hypothesis was put forward by Yukawa and it was not discovered until 1947 when Powell and his colleagues detected a particle, now known as the *pi-meson* or *pion* (denoted by π) during their studies of cosmic rays, which satisfied all of the requirements. Cosmic rays consist of a variety of particles falling on the earth from outer space which they detected by exposing photographic plates at high altitudes. Particles passing through the plates disturb the photographic emulsion and leave 'tracks' whose nature enables the particle mass to be estimated. The mass turned out to be around 270 times the electron mass and so Yukawa's idea was established. The pion was also found later to have spin 0. In figure 8.3(b) a representation of the pion exchange process is shown where the pion is denoted by the dotted line. Diagrams such as those in figure 8.3 are generally referred to as *Feynman diagrams*.

In the same way that the strength of the electric force is determined by the charge e, so the strength of the nuclear force is denoted by a symbol g which is referred to as a *coupling constant* since it measures the strength with which the pion is coupled to the nucleon. To account for the strength of the nuclear force g^2 needs to be around $100e^2$. This means that this interaction is strong compared with the electromagnetic interaction and it is one manifestation of what has already been referred to as the *strong interaction*. Many more exchanged particles, generally referred to as *mesons*, are now known to contribute to the nuclear force. Some, including pions, carry electric charges; for example pions exist with no charge (π^0), with charge $+e$ (π^+) and with charge $-e$ (π^-), all three particles having essentially the same mass. As a result a neutron and a proton interacting with each other can exchange their identities by exchanging, for example, a π^+. Such a charge exchange process is illustrated in figure 8.3(c) and leads to the frequently observed phenomenon of neutrons and protons 'changing places' when scattering one another.

In terms of these many different and varied processes, a very full understanding of the complex nature of the nuclear force has now been obtained.

8.4 Nuclear Models

The extreme complexity of the nuclear force coupled with the fact that nuclei can contain up to 250 or more nucleons means that it is very difficult to develop an exact theoretical treatment of nuclear structure. The approach during the last 50 years has therefore been to devise conceptual models of the nucleus which are simpler to deal with but which are reasonably close to its physical structure and which enable quite detailed understanding of nuclear behaviour to be achieved. Such models tend to focus on different aspects of nuclear structure and on different groups of nuclei.

One of the simplest and oldest models, which accounts for the general aspects of nuclear masses and binding energies, is known as the *liquid drop model*. This model is based on the similarity between a drop of liquid and an atomic nucleus. In a liquid drop the component atoms or molecules are held together by a force which is vaguely similar to, although of very different range and strength (see the previous section), to the nuclear force. For example, in both systems there is a net inward force on their components at the surface which results in a 'surface tension' effect (see section 3.6). This constrains drops and nuclei to be roughly spherical in shape. By treating the nucleus as a liquid drop and including this surface tension effect together with an additional force that does not occur in a liquid drop, namely the electric repulsion between the protons, it becomes possible to account in reasonable detail for the shape of the B/A curve in figure 8.2.

Such a model is essentially classical, however, and therefore cannot help with understanding the energy levels in a nucleus. This was first achieved by the introduction of the *nuclear shell model* which is analogous in many ways to an atomic-like description of the nucleus. Focusing attention on a single neutron as it moves through the nucleus there is, on average, no net force on it due to other nucleons whilst in the middle of the nucleus since they surround it fairly uniformly. It simply experiences a roughly constant potential energy due to them as shown in figure 8.4. However, as it moves towards the nuclear surface, there will be an increasing attractive inward force (cf surface tension—

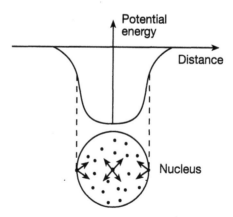

Figure 8.4: *The average potential energy experienced by a neutron in a nucleus.*

section 3.6) leading to a rise in the potential energy. This eventually reduces to zero as the neutron moves away from the nucleus and out of the range of the nuclear force. The neutron is thus confined in a potential well similar in general features to those discussed in section 5.4. Protons are similarly confined but the wells have a slightly different shape at the edges due to the additional electric force they experience.

In such a well a series of different quantum energy levels will be available for the nucleons to occupy, each level being characterized by several quantum numbers in a similar fashion to the energy levels available to an electron in an atom (see section 5.5). Because of the Pauli exclusion principle, each level will only be able to accommodate a certain number of neutrons and protons and, when full, we speak of *filled* shells as in the analogous atomic situation. These 'fillings' occur when the number of protons (Z) and the number of neutrons (N) take the following values (referred to as *magic numbers*).

$$Z = 2, 8, 20, 28, 50, 82$$

$$N = 2, 8, 20, 28, 50, 82, 126.$$

Nuclei with filled shells are particularly stable. Examples are helium $(Z = N = 2)$, oxygen $(Z = N = 8)$ and lead $(Z = 82, N = 126)$. These nuclei are in this respect somewhat like atoms of the inert gases (see section 5.5). The energy levels and other (e.g. electromagnetic) properties of nuclei containing a few nucleons in excess of, or less than, magic numbers can then be very readily accounted for in considerable detail in terms of the behaviour of these 'loose' nucleons or holes—a nucleon missing from a shell corresponds to a hole in the shell. These interact with each other through, as it were, the remnants of the nuclear force not included in the average potential well. However when there are many loose nucleons the situation becomes extremely complicated and another approach to nuclear structure was developed—the *collective model.*

Loose nucleons tend to alter the nuclear shape and with many of them the nucleus can become quite distorted. If there are not too

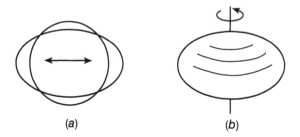

(a)   (b)

Figure 8.5: *Nuclear collective motion: (a) vibrational; (b) rotational.*

many it is found that the whole nucleus oscillates in shape, for example, backwards and forwards from spherical (soccer ball) to ellipsoidal (American football) shape (see figure 8.5(a)). These oscillations are, of course, quantized and lead to a series of equally spaced oscillator energy levels (see figure 5.3(b)), many examples of which have been observed. Another form of collective oscillation which also occurs is when the neutrons and protons oscillate *en masse* in opposite directions to each other.

When the number of loose nucleons is very high nuclei become permamently deformed; in some extreme cases the length and width for an American-football-like shape are in the ratio 2:1. For such nuclei it is natural for them to rotate as shown in figure 8.5(b), leading to a series of quantized 'rotational' energy levels. Angular momenta as large as 60 $\hbar$ have been observed. For much larger values the nucleus will fly apart because of the centrifugal force.

These different nuclear models have formed the basis for understanding many nuclear properties. They must all, in the deepest sense, be related to each other and much very complicated theoretical work is in progress now to clarify the situation.

8.5 Nuclear Reactions

A nuclear reaction is a process in which a 'target' nucleus is knocked about, and possibly broken up, by bombarding it with a beam of nucleons or other nuclei. A beam of protons or light nuclei (accompanied by some of their atomic electrons in the form of ions) is obtained by subjecting them to a large electric potential difference (see section 4.2), which, because of their electric charge, accelerates them to a high energy. *Accelerators* are either linear, in which a high potential difference is maintained between two points, or circular, in which the particles are constrained by magnetic fields to move in circular orbits and are continually accelerated under the influence of oscillating electric fields. Neutrons, having no electric charge, cannot be accelerated in this way and beams have to be produced using an intermediate nuclear reaction as in the Chadwick experiment described in section 8.1. The products of the nuclear reaction are investigated using *detectors* or *counters* such as scintillation counters (developments of the process used by Geiger and Marsden mentioned in section 5.1), ionization detectors (see section 8.1) and semiconductor detectors (in which particles excite electrons in a semiconductor into the conduction band, as in solar cells mentioned in section 6.3, so giving an electric signal).

Nuclear reaction processes take essentially two forms. First, there are what are called *direct reactions* in which the disturbance of the nucleus by the bombarding particle is not very great. The particle may simply be scattered from the target nucleus without affecting it at all; it may knock a nucleon in the nucleus into a higher-energy state, leaving the nucleus excited, or even knock it out; it may 'pick up' a nucleon from the nucleus and carry it away or be 'stripped' of one of its own component nucleons, leaving a different target nucleus.

Second, there are *compound nucleus reactions* in which the bombarding particle gives up all its energy to the target nucleus and cannot escape thus forming a new (compound) nucleus. Of course, this compound nucleus has a lot of energy and eventually after· a lot of thrashing about and after a relatively long time (perhaps a million or more times longer than the time which would be taken for the bombarding particle simply to pass through the nucleus) this energy is concentrated onto one or more nucleons again which then escape, or the nucleus may get rid of the energy in the form of a photon. Whereas a direct reaction is essentially a continuous process, a compound nucleus reaction has two stages—the formation of the compound nucleus and then, after a while, the emission of one or more nucleons or a photon. The compound nucleus process is also characterized by showing 'resonant' behaviour. That is, the probability of it happening is much greater at certain values of the energy of the bombarding particle. This happens when that energy is such that the resultant energy of the compound nucleus coincides with one of its energy levels (see section 6.2). Detailed study of these different processes is the preoccupation of many nuclear physicists and gives considerable information about the properties and distribution of nuclear energy levels.

There are two other nuclear reaction processes which are of particular interest, namely *fusion* and *fission*. A fusion process is one in which two very light nuclei coalesce and energy is released. The best known of these is the fusion process responsible for 'hydrogen burning' in stars (see section 10.4). In this process a series of nuclear reactions takes place starting with two protons (hydrogen nuclei) joining together to form the heavy hydrogen nucleus (^{2}H) consisting of one proton and one neutron. To achieve this one proton has to convert to a neutron and this happens through the *beta-decay* process which will be discussed in the next section. The heavy hydrogen nucleus then captures another proton, forming the nucleus of an isotope of helium (two protons and one neutron, ^{3}He) and finally two of these nuclei join together and give up two protons so leaving the very stable helium nucleus (^{4}He) consisting of two protons and two neutrons. The essential feature of these processes is that the final helium nucleus formed is very stable (see remarks in the previous section) and a great deal of energy is released in its formation; conversely much energy is needed to knock it to pieces. The energy released in fusing very light nuclei together to form heavier nuclei follows in general terms from figure 8.2, in which it can be seen that the binding energy per nucleon moving from very light to heavier nuclei increases, so leading to a release of energy in the fusion process. The magnitude of the energy released in such processes is of the order one million or more times the energy released in a chemical process.

For decades physicists have been trying to achieve fusion on a commercial scale as a source of power and are gradually approaching this target. There are two approaches to the problem. One is to confine deuterium (^{2}H) and tritium (^{3}H) nuclei, together with loose electrons, in the form of a *plasma* (an assembly of ionized atoms and electrons) by powerful magnetic fields at temperatures around 10^8 K. At this temperature the thermal kinetic energies of the nuclei are such that they can overcome the electric repulsion between themselves and fuse together. The other approach is to use *inertial fusion* in which a pellet of ^{2}H and ^{3}H is violently compressed using, for example, a laser beam or, as in a hydrogen bomb, a fission explosion (see below).

Turning now to nuclear fission, if a very heavy nucleus splits into two lighter nuclei, the curve in figure 8.2 indicates again that the binding energy per nucleon is greater in the fission products than in the heavy nucleus so energy is released in the process. Spontaneous fission, is a very rare process, but fission can be induced by bombarding a heavy nucleus such as uranium with neutrons: a neutron is captured by the nucleus and the resultant compound nucleus becomes so unstable that it readily splits apart. What is of extreme importance is that in the fission process a few spare neutrons are also let loose from the fissioning nucleus and these can then provoke fission in neighbouring nuclei and so on and so on. This is the origin of a *chain reaction* in which, given a sufficiently large collection (*critical mass*) of closely packed uranium nuclei, one loose nucleon sets in motion an escalating series of fission processes and, hence, a nuclear explosion. Such an explosion is controlled in a *nuclear reactor* by using *control rods* (usually of boron or cadmium), which are very good absorbers of neutrons and so limit the extent to which fission takes place. The energy released, again of the order a million or more times chemical energies, is then used to heat a circulating coolant to provide steam to drive a turbine for electric power generation.

8.6 Radioactivity

Towards the end of the 19th century various experiments indicated that different types of hitherto unknown radiation were emitted by heavy atoms. In 1896 Becquerel discovered that uranium compounds emitted a form of radiation (beta- or β-radiation) which could be deflected by a magnetic field. This implied that the radiation consisted of a stream of charged particles and these were subsequently identified to be electrons. Subsequently, positron (see section 7.6) emission was also observed. Then, in 1897–8, Pierre and Marie Curie found another form of much less penetrating charged particle radiation (alpha- or α-radiation) which, as mentioned in section 5.1, is now known to consist of ionized helium, i.e. helium nuclei. Finally, in 1900, Villard identified a third form of highly penetrating radiation (gamma- or γ-radiation) which was undeflected by a magnetic field and turned out to be akin to x-rays (see section 4.4) but of much shorter wavelength.

Each of these radioactive decay processes happen spontaneously and can be understood to take place when an atomic nucleus is in an excited or unstable state and is able to release energy. Natural occurring radioactivity arises in very heavy nuclei such as uranium and radium which, since they contain many protons all repelling each other, are liable to disintegrate into more stable structures. Artificial radioactive nuclei are produced as the products of nuclear reactions.

Quantum mechanics does not allow us to predict exactly when such a radioactive decay will take place, only the probability of it happening in a given time. The standard unit of activity of a radioactive substance is the *becquerel* which corresponds to one decay per second. All radioactive decays are then characterized by a time known as the *mean life*, which is the average lifetime of a nucleus before it decays. Mean lives can vary from very high values of the order of 10^{10} years for naturally occurring alpha-emitters through to 10^{-15} s for some gamma-emitters. Let us now consider the three forms of radioactivity in turn.

Alpha-decay (α-decay). We have seen that the helium nucleus is a very stable structure and that, correspondingly, when it is assembled by bringing two neutrons and two protons together a great deal of energy is released. This means that, since nuclear systems are always seeking to be in the lowest possible energy state, one way of achieving this for an excited or very heavy nucleus may be by ejecting two protons and two neutrons in the form of a helium nucleus—an α-particle, the energy released in forming the helium nucleus being used to enable it to escape from the 'parent' nucleus. In this escape the α-particle has to penetrate a potential barrier as described in section 5.4. To conserve energy, the α-particle and the 'daughter' nucleus in such a decay carry away a definite amount of kinetic energy between them equal to the differerence in mass–energy (remember $E = mc^2$) between the parent nucleus and the daughter nucleus plus the helium nucleus. This is observed experimentally and in the foregoing terms the process of α-decay is now well understood.

Beta-decay (β-decay). This similarly involves the emission of a charged particle—an electron (e^-) or a positron (e^+). This happens when there are respectively too many neutrons or too many protons in a nucleus for it to be stable. Consequently, a neutron changes into a proton, emitting an electron, or a proton into a neutron, emitting a positron. Given that nuclei have definite energies it would be expected that, as with α-decay, the β-particle would also have a definite energy. It was, therefore, a great surprise to find that in any β-decay process the β-particles were emitted with a wide spread of kinetic energies from zero up to the maximum available. The solution to this problem was suggested by Pauli in 1930 who proposed that in β-decay *two* particles were emitted so that the definite energy released was shared between them. The electron or positron could then have all possible values of the energy up to the maximum as observed the rest being taken away by the second particle. Since all the electric charge released in β-decay is carried away by the electron or positron it follows that the new particle is electrically neutral, like a neutron. It also has a very small, if not zero, mass since electrons and positrons are observed sometimes to have essentially all the energy released in the decay—there is no spare energy for the new particle to have a significant mass. For these reasons the new particle was seen as a 'little' neutron and was called a *neutrino* (denoted by v_e). It also became clear that, if angular momentum is to be conserved in β-decay then, like an electron, it must have spin $\frac{1}{2}$. Similarly, it also has a corresponding antiparticle called an *antineutrino* (denoted by $\bar{v}_e$).

The introduction of the neutrino enabled the β-decay process to be well understood in terms of two basic processes—a neutron converting into a proton with the creation of an electron and an antineutrino ($n \rightarrow p + e^- + \bar{\nu}_e$) or a proton converting into a neutron with the creation of a positron and a neutrino ($p \rightarrow n + e^+ + \nu_e$). Since a neutron is slightly heavier than a proton, the first process can also happen to a free neutron and its mean life is measured to be around 15 minutes, but the second process can only happen when the proton is embedded in a nucleus, which, if it is radioactive, provides the energy to create the positron and the neutrino. However, it was not until 1953 that the neutrino (actually it was the antineutrino) was detected experimentally (by Reines and Cowan) in the vicinity of a nuclear reactor. This was possible because near a reactor there are many free neutrons (see the previous section) and, as we have just seen, they decay reasonably quickly into a proton, an electron and an antineutrino. The antineutrino was detected by letting it be captured by a proton, so converting the proton into a neutron and a positron ($p + \bar{\nu}_e \rightarrow n + e^+$), the latter being easily observed. This capture process is known as *inverse β-decay*. Here it should be mentioned that sometimes a nucleus, instead of emitting a positron, captures an orbiting atomic electron, so converting a proton into a neutron and emitting a neutrino ($p + e^- \rightarrow n + \nu_e$).*

Gamma-decay (γ-decay). This process can easily be understood as the nuclear equivalent of the emission of photons from an atom. Like an atom the nucleus has a series of energy levels and if the nucleus is excited to a higher one, for example in a nuclear reaction or following radioactive decay, it will make a series of jumps down to the lower ones finally ending up in the 'ground' state. Each jump involves the release of energy and this is carried away as a photon—the γ-radiation. Since nuclear energies are on a much larger scale than atomic energies (millions of electron volts rather than electron volts) it follows that the corresponding frequencies of the emitted photons (energy $h\nu$) are some millions of times larger than those of visible radiation and the wavelength of the radiation some millions of times smaller (even smaller than for x-rays—see table 4.1) as observed.

* Here it should be noted that when particles are switched from one side of a reaction arrow ($\rightarrow$) to the other, they are replaced by their antiparticles.

8.7 Nuclear Physics—a Few Remarks

The study of nuclear physics led to the realization that in order to understand the behaviour of atomic nuclei it was necessary to postulate the existence of elementary particles other than the proton (p) and the electron (e). In particular, as we have seen, the existence of the neutron (n) was established. Further, to account for the nuclear force, quite new types of particle had to be introduced—pions ($\pi^{+,-,0}$) and other mesons. Then, in order to understand the beta-decay process, a further particle was postulated—the neutrino (ν_e) and the corresponding antineutrino ($\bar{\nu}_e$).

Over the years very detailed and painstaking investigation of nuclear processes and properties have given vital information about the interactions involving these new particles. For example, and as will be discussed fully in section 9.4, the study of beta-decay showed that the interaction responsible for this process violated one of the fundamental symmetries believed to govern all physical processes. This symmetry (referred to technically as *parity conservation*) postulated that the mirror image of any basic physical process could also occur in the real world. Currently, studies of beta-decay are, among other things, aimed at clarifying the important issue of whether or not the neutrino has a finite mass.

Turning to nuclear forces, a great deal of information about their detailed nature has come from studying the way in which nucleons are scattered from each other using accelerators and also from the properties of very simple nuclei containing two or three nucleons only. In the case of three-nucleon nuclei it has also emerged that three-body forces exist, which are such that the force between two nucleons is altered when a third nucleon is nearby. Returning to the tennis analogy to pion exchange mentioned in section 8.3, it is rather like the interaction between two tennis players when a third starts to interfere. In this 'exchange' context, an additional magnetic effect has also been observed in simple nuclei. Since a meson such as the pion can carry electric charge, there is a resultant electric current as it is exchanged between nucleons which, in turn, produces a magnetic field. This contributes to the magnetic properties of a nucleus (called a *magnetic exchange effect*).

The above are just a few examples of the ways in which nuclear physics is able to throw light on fundamental elementary particle processes. Work of this kind continues as well as studies of what might be called 'exotic' nuclei, for example nuclei which have an excessive number of neutrons compared with the number of protons and which, as a result, have a 'halo' of neutrons extending well beyond the potential well confining the nucleons (see figure 8.4). Study of such nuclei can give information about 'neutron matter', which can help with our understanding of neutron stars (see section 10.5). Nuclear physics continues to be an exciting part of physics.

8.8 Moving Forward

The nucleus, as was emphasized at the outset of this chapter, is a complex entity with which to deal, not least because of the complicated nature of the nuclear force. However, it should be clear that its behaviour is now understood reasonably well even though it is difficult to work out some of the very fine details. As has been mentioned in the last section, in achieving this understanding of its structure and its properties, we have had to deal with a number of 'elementary particles' It turns out that these are only a few of literally many hundreds of such particles that are now known to exist and in the next chapter an outline is given of the way in which this panoply of particles can be understood as an ordered manifestation of a much smaller array of even more basic particles.

CHAPTER 9

THE FUNDAMENTAL CONSTITUENTS OF MATTER

Elementary Particles and their Interactions

9.1 The Classification of Elementary Particles

In section 8.3 the electromagnetic interaction was described in terms of the exchange of photons between particles and the nuclear force in terms of the exchange of mesons between nucleons. These processes involve the creation and annihilation of the exchanged photon or meson by the interacting particles and, as already mentioned, the strength of the interaction, which is related to the probability of creation and annihilation processes, is measured by the square of the relevant coupling constant. It is clear from the relative size of these quantities that the nuclear interaction is around 100 times stronger than the *electromagnetic interaction*. It is an example of what is generically called the *strong interaction*, which, as we shall see, is experienced by a great many particles. We also encountered the beta-decay creation and annihilation processes involving nucleons as well as electrons, neutrinos and their corresponding antiparticles. The probability of such processes happening, as indicated by their mean lifetimes, indicate that the coupling constant responsible for the creation of, say, electrons and antineutrinos in beta-decay is very small indeed—many orders of magnitude less than the electromagnetic coupling constant. For this reason, beta-decay and many other similar processes are said to derive from the *weak interaction*.

In the light of these comments it is now possible to classify the many different elementary particles met within the physical world in terms of their masses, spins and the interactions they experience. Very broad classifications are as follows.

Photon. *This is the massless quantized manifestation of an electromagnetic field and only experiences the electromagnetic interaction. It has spin 1 and, as mentioned in section 8.3, is an example of what is referred to as a gauge boson.*

Leptons. *The collective name for the very small number of particles such as electrons and neutrinos which only experience the weak and, if electrically charged, the electromagnetic interaction. They all have spin $\frac{1}{2}$.*

Hadrons. *The collective name for particles which experience the strong interaction as well as the weak and, usually, electromagnetic interactions. They are divided into two groups.*

    Mesons. *These are those hadrons, like the pion, which have integer spin (usually 0 or 1).*

    Baryons. *These are those hadrons, like nucleons, which have half-integer spin (usually $\frac{1}{2}$ or $\frac{3}{2}$).*

Apart from those elementary particles which have been encountered so far, it is now known that there are literally hundreds more which fall into one or other of the above categories. These have largely been discovered using high-energy accelerators. In section 8.5 the basic mechanisms for accelerating charged elementary particles were mentioned. Over the last 50 years accelerators have been built producing beams of particles with higher and higher energies. These energies are measured in electron volts (see Glossary) and, to put accelerator energies into perspective, remember that the energy needed to knock a typical nucleon from a nucleus is around 8 MeV (see figure 8.2). Accelerators used in nuclear physics usually have energies around 20–40 MeV, sufficient to knock several nucleons out of a nucleus, but in investigating elementary particles much higher energies are used. For example, the Super Proton Synchrotron at CERN (European Centre of Nuclear Research in Geneva) is a circular machine having a circumference around 6 km and producing protons of energy 450,000 MeV (or 450 GeV). In such an accelerator the protons are accelerated by radio-frequency electromagnetic fields and kept in essentially fixed orbits by powerful magnetic fields whose value increases as bunches of protons continually accelerate. The Large Electron-Positron Collider (LEP) at CERN accelerates electrons and positrons in opposite directions in circular orbits with a circumference of 27 km. They then collide with each other, the energy of the collision being up to 190 GeV. The tunnel of LEP is also to accommodate what is known as the Large Hadron Collider (LHC) in which protons will be accelerated in opposite directions and collide with each other at energies which will eventually reach around 14,000 GeV. Linear machines are also in use, for example the Stanford Linear Accelerator (SLAC) in the USA, in which electrons are accelerated over a distance of around 3 km to energies of 50 GeV.

When charged particles are accelerated in, for example, circular machines they continually emit electromagnetic radiation (section 4.4). This is known as *synchrotron radiation* and some machines are specifically designed to produce this radiation as an intense source of x-rays for use in experimental studies in many fields of physics.

High-energy particle accelerators are massive undertakings and can generally only be financed on an international basis. Their immense energies do mean that it is possible to create particles of masses much larger than the heaviest particles we have encountered, namely the neutron and proton, whose mass energy ($E = mc^2$) is around 1 GeV. Hundreds of new particles have been created in this way and are found to be very unstable. They decay very quickly and it is for this reason that they have not appeared directly in the physical processes we have dealt with so far. They have been identified using detectors of the type already mentioned in section 8.5 together with other devices such as, for example, *bubble chambers* and, more recently, *wire chambers*. The former are large chambers containing low-temperature liquified gases such as hydrogen near to boiling which are such that, when a charged particle passes through, local boiling takes place and bubbles along its track are formed. The way in which such tracks bend in a magnetic field and also their lengths then enable the mass and mean life of the particle to be deduced. Wire chambers contain positively and negatively charged wires immersed in an appropriate gas.

Charged particles entering the chamber knock electrons out of the gas atoms and the resultant ions then go on to knock further electrons out of other atoms. The electrons in the resulting pulse are attracted towards a positively charged wire, giving a large signal. Analysis of such signals then enables the position and momentum of the incoming particles to be measured very accurately.

To give some impression of the various particles which have now been identified, table 9.1 lists just a few of them in each of the categories referred to earlier together with their names, symbols (frequently Greek letters), approximate masses (as multiples of the electron mass m_e), approximate mean lives (in seconds (s)) and the most probable products of the decay. Also, for hadrons, one other quantity known as *strangeness* is given; this will be explained in the next section. It should also be remembered that corresponding to each particle is an *antiparticle* (e.g. see section 7.6) having the same mass and spin as the particle and with an opposite and equal value for quantities such as electric charge. Antiparticles are denoted by adding an overbar to the particle symbol: for example, an antiproton is denoted by $\bar{p}$.

9.2 Intrinsic Particle Properties and Conservation Laws

As we have discussed elementary particles so far they have been characterized by essentially three quantities—their masses, electric charges and spins. These quantities are important since, as we have seen, they all quantities which feature in conservation laws. The three relevant laws refer to the conservation of mass–energy (see section 7.5), the conservation of charge (see section 4.1) and the conservation of angular momentum (see section 2.3). As is to be expected, all elementary particle processes obey these conservation laws as well as the law of conservation of momentum (section 2.3), but it is found in studying the way in which elementary particles interact that other simple conservation laws must be in operation. For example, in any process it is found that the number of baryons less the number of antibaryons at the outset of a process is the same as at the end. This can be

Table 9.1: *A few elementary particles.*

Leptons (spin $\frac{1}{2}$)			
Name (symbol)	Mass (m_e)	Mean life (s)	Decay products
Electron (e⁻)	1	stable	none
Electron neutrino (ν_e)	$< 2 \times 10^{-5}$	stable	none
Muon (μ^-)	207	2×10^{-6}	$e^- + \bar{\nu}_e + \nu_\mu$
Muon neutrino (ν_μ)	< 0.5	stable	none
Tau (τ^-)	3500	3×10^{-13}	$\mu^- + \bar{\nu}_\mu + \nu_e$
Tau neutrino (ν_τ)	?	?	none

Baryons (spin $\frac{1}{2}$)				
Name (symbol)	Mass (m_e)	Charge	Strange-ness	Mean life (s)	Decay products
---	---	---	---	---	---
Proton (p)	1836	$+e$	0	stable	—
Neutron (n)	1839	0	0	900	$p + e^- + \bar{\nu}_e$
Lambda (Λ^0)	2183	0	-1	2.6×10^{-10}	$p + \pi^-$
Sigma plus (Σ^+)	2328	$+e$	-1	0.8×10^{-10}	$p + \pi^0$
Sigma nought (Σ^0)	2334	0	-1	7.4×10^{-20}	$\Lambda^0 + \gamma$
Sigma minus (Σ^-)	2343	$-e$	-1	1.5×10^{-10}	$n + \pi^-$
Xi nought (Ξ^0)	2573	0	-2	2.9×10^{-10}	$\Lambda^0 + \pi^0$
Xi minus (Ξ^-)	2586	$-e$	-2	1.6×10^{-10}	$\Lambda^0 + \pi^-$

Mesons (spin 0)					
Name (symbol)	Mass (m_e)	Charge	Strange-ness	Mean life (s)	Decay products
Pion plus (π^+)	273	$+e$	0	2.6×10^{-8}	$\mu^+ + \nu_\mu$
Pion nought (π^0)	264	0	0	0.8×10^{-16}	$\gamma + \gamma$
Pion minus (π^-)	273	$-e$	0	2.6×10^{-8}	$\mu^- + \bar{\nu}_\mu$
Eta nought (η^0)	549	0	0	1×10^{-18}	$\gamma + \gamma$
Kaon plus (K^+)	966	$+e$	$+1$	1.2×10^{-8}	$\mu^+ + \nu_\mu$
Kaon nought (K^0)	974	0	$+1$	0.9×10^{-10} $(K_S^0)^*$	$\pi^+ + \pi^-$
				5.2×10^{-8} $(K_L^0)^*$	$\pi^+ + e^- + \bar{\nu}_e$

* As will be explained later the K^0 can exist in two forms—one lives for a relatively short time before decaying (suffix S) and one lives for a significantly longer time (suffix L).

formulated quantitatively by giving every baryon a *baryon number* (denoted by *B*) equal to +1, every antibaryon $B = -1$ and all other particles $B = 0$ and then requiring that the total baryon number in any process is conserved—the law of *baryon conservation*. This is exemplified in the following processes—only one of which can take place. The other, although otherwise allowed, cannot happen because baryon number is not conserved.

$$p + n \rightarrow p + p + \pi^-$$
$$B = 1 \quad 1 \quad 1 \quad 1 \quad 0$$
(*allowed* since the total value of *B* is 2 initially and finally)

$$p + n \rightarrow p + \bar{p} + \pi^+$$
$$B = 1 \quad 1 \rightarrow 1 \quad -1 \quad 0$$
(*disallowed* since $B = 2$ initially and $B = 0$ finally)

where the symbol $\bar{p}$ denotes an antiproton with, of course, a negative electric charge.

There is a similar conservation law for leptons and their antiparticles. In table 9.1 it will be seen that besides the electron and the neutrino that we have met in nuclear beta-decay, there are two further pairs of leptons—the muon together with its neutrino and the tau together with its neutrino. It is found that only those processes take place in which the three different brands of lepton are separately conserved. For example, if a muon is emitted or absorbed then its associated neutrino (or antineutrino) will also be involved.

Another conservation law of this kind and of great importance is *strangeness conservation* which only applies, however, to processes resulting from the strong and electromagnetic interactions. The introduction of the concept of 'strangeness' arose in the early 1950s to account for the fact that some of the new particles being produced at that time in accelerator experiments could only be created in association with each other. For example the K^0 could be produced in association with a Λ^0 and not in association with a neutron. The first of the following processes happens, but the second does not even though none of the conservation laws discussed so far is violated:

$$\pi^- + p \to \Lambda^0 + K^0$$
$$S = 0 \quad 0 \quad -1 \quad +1$$

$$\pi^- + p \to n + K^0$$
$$S = 0 \quad 0 \quad 0 \quad +1.$$

The solution to this strange behaviour was to introduce a conservation law that *was* violated by the second process. To this end a number called *strangeness* (denoted by S) was allocated to all strongly interacting particles (with the opposite sign for the corresponding antiparticle) as in table 9.1. Since these 'strange' particles were produced prolifically the production process must involve the strong interaction and it was therefore proposed that it conserved strangeness as in the first of the above processes, but not the second. It will be noted in the table that the Ξ-particle has $S = -2$ and, therefore, can only be produced in association with, for example, *two* K^0-particles (each having $S = +1$). However, it will be seen from the table that strangeness is not conserved when strange particles decay—it changes by 1. Such decays result from the weak interaction, which therefore does not conserve strangeness.

Conservation laws of the type that we have been discussing—baryon, lepton and strangeness conservation are, like charge conservation, simple additive laws. The sum of the initial quantities (e.g. B or S) must be equal to the sum of the final quantities. It should be stressed here that, although quantities such as charge, baryon number and strangeness were initially introduced to help codify what can and cannot happen, when detailed theories of these different processes are formulated, the conservation laws arise naturally by requiring the theories to be *invariant* under certain mathematical transformations. A simple example of invariance in classical physics is that the laws of conservation of momentum and of energy follow from the requirement that the theory of classical mechanics remains the same (i.e. is invariant) whatever position in space and time it is referred to. If it were not then the nature of motion would change from place to place and from time to time! It clearly does not. Invariance under more complicated and less intuitively obvious transformations lead on to the other conservation laws.

9.3 Understanding the Nature of Hadrons

In just the same way that the atomic nuclei can be understood in terms of different arrangements of two basic particles—the proton and neutron—so it might be hoped that the many different elementary particles could be similarly understood in terms of a few basic components. Such a step forward in our understanding did take place in the early 1960s due mainly to the brilliant ideas of Murray Gell-Mann (Nobel Laureate, 1969). He suggested that all hadrons were formed from different combinations of three basic entities which he called *quarks* ('...three quarks for Muster Mark...'—James Joyce in Finnegan's Wake), the quarks being bound together in the particle by an interquark force. This proposal was substantiated by experiments in which electrons from SLAC (see section 9.1) were scattered by protons. Just as the scattering of alpha-particles by atoms indicated the presence of atomic nuclei (see section 5.1) so the scattering of these high-energy electrons indicated the presence of three pointlike entities in the interior of protons. The three quarks, which have corresponding antiquarks, were given the individual names, symbols and properties listed in table 9.2. Here, for the first time we meet particles with one-third-integral charges. The reason for this is that if quarks are to be components of neutrons and protons then the simplest possibility is to have three of them. Two of them would not give the right spin (in quantum mechanics two spin-$\frac{1}{2}$ particles can only lead to an integer spin, whereas neutrons and protons have spin $\frac{1}{2}$). So it was postulated that *all* baryons consisted of different combinations of these three quarks.

Table 9.2: *Quarks and their properties.*

Name	Symbol	Spin	Charge	Baryon number	Strangeness
Up	u	$\frac{1}{2}$	$+\frac{2}{3}e$	$\frac{1}{3}$	0
Down	d	$\frac{1}{2}$	$-\frac{1}{3}e$	$\frac{1}{3}$	0
Strange	s	$\frac{1}{2}$	$-\frac{1}{3}e$	$\frac{1}{3}$	-1

Examples of the different compositions are given below for the proton (p), neutron (n), sigma minus (Σ^-) and Xi minus (Ξ^-).

$$
\begin{array}{ll}
\text{p} & \text{consists of } \; u + u + d \\
\text{n} & \text{consists of } \; u + d + d \\
\Sigma^- & \text{consists of } \; d + d + s \\
\Xi^- & \text{consists of } \; d + s + s.
\end{array}
$$

Simple arithmetic will confirm that each particle has the charge, baryon number and strangeness given in table 9.1.

Similarly, mesons can be understood as consisting of one quark and one antiquark. As just pointed out the combination of two spin-$\frac{1}{2}$ particles gives an integer spin as required for mesons. Further, since antiquarks have baryon number $B = -\frac{1}{3}$, it follows that the total baryon number for a meson is 0, also as required. Examples of the quark structure of a few mesons—pion plus (π^+), kaon plus (K^+) and kaon nought (K^0)—are given below.

$$
\begin{array}{ll}
\pi^+ & \text{consists of } \; u + \bar{d} \\
K^+ & \text{consists of } \; u + \bar{s} \\
K^0 & \text{consists of } \; d + \bar{s}.
\end{array}
$$

It must be stressed here that the three quarks were not introduced on an *ad hoc* basis but as the basic states of a very elegant mathematical theory of elementary particles known symbolically as SU(3) (standing for 'the special unitary group in three dimensions'). This theory could account for the obvious grouping of particles such as the neutron and proton and the three types of π-meson, members of these two groups having virtually identical properties as far as the strong interaction is concerned. For example, members of each group have approximately the same mass and only differ in their electric charge. It also predicted larger groupings. For example, the eight baryons listed in table 9.1 constitute one such group. The particles in this group do differ somewhat in their masses and so SU(3) symmetry is obviously not exact. It follows that the three basic quarks must have different masses. For example, the u- and d-quarks will each have masses around one-third of the mass of a neutron or proton, since each contains three quarks, whereas the s-quark must be heavier in order to account for the higher masses of the various ·'strange' baryons. An even larger group containing ten baryons with spin $\frac{3}{2}$ was also predicted. One of these baryons—the Ω^-—contained three s-quarks and it was a great triumph for the theory when the Ω^- was detected in 1964 with the mass and properties predicted.

Since then, however, as accelerator energies have increased, new particles have been discovered which cannot be accounted for in terms of the three basic quarks discussed so far. The theory had to be extended and three further heavier quarks, all with one-third-integral charges, were introduced. They were allocated 'flavours' (analagous to strangeness) referred to as *charm* (*c*), *bottom* (*b*) and *top* (*t*) having the values +1, −1 and +1 respectively, the three original quarks (u, d, s) having the value 0. These flavours, like strangeness, are conserved additively in strong and electromagnetic processes. In terms of theories including these six basic quarks much understanding of elementary particle processes has been achieved.

There is another point to consider. Quarks have spin $\frac{1}{2}$ and therefore Pauli's exclusion principle (see section 5.5) applies when considering the quantum states of two or more identical quarks. Consider, for example, the Ω^- referred to above, consisting of three s-quarks, which, since it has spin $\frac{3}{2}$, must all have their spins pointing in the same direction—they are all in the same state and this is not allowed by the exclusion principle. To solve this problem it was suggested that quarks have a sort of electric charge, which was called *colour charge*, and that there were three varieties of colour charge—*red, yellow* and *blue*. In the Ω^- each s-quark was then postulated to have a different 'colour' so that they were no longer identical and the exclusion principle was not violated. In this 'colour' language, remembering that mixing the three primary colours gives white, they can be said to be in a 'white' state. As the theory was developed it transpired that all elementary particles are 'white'; baryons all contain three differently coloured quarks and mesons all contain quarks and antiquarks whose colours exactly cancel (just as the charges of a particle and its antiparticle cancel). Here it must be stressed again that the three colours are purely a short-hand terminology for quantities in the theory of strong interactions of quarks which play a similar role to electric charge in electromagnetic theory. They are nothing to do with the real colours of everyday life.

If this theory of particle structure is correct, it is surprising that no isolated quark has ever been observed. Whilst electrons can be knocked out of an atom, or neutrons and protons out of a nucleus, it proves impossible to knock a quark out of an elementary particle. This implies that the inter-quark force has some unusual features. In particular it must be rather like the force experienced between two entities joined together by a piece of elastic or a spring; the further they are separated the stronger the force pulling them together. This is completely opposite to electromagnetic and strong interaction forces which become weaker as two interacting particles are separated. This peculiar property of the force is referred to as *confinement* and arises naturally in a theory which has been formulated known as *quantum chromodynamics* (QCD), which attributes the force to the exchange of massless entities between quarks and whose existence has been well confirmed experimentally. Like photons, they are *gauge bosons* having spin 1 and are referred to picturesquely as *gluons*. The exchange process is shown diagrammatically in figure 9.1 where the gluon is represented by a 'coiled spring' reflecting the nature of the force it produces. Comparison should be made with the photon and pion exchange processes shown in figure 8.3.

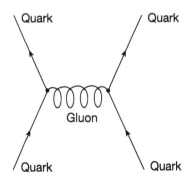

Figure 9.1: *Gluon exchange between quarks.*

Just as the strength of the electromagnetic force due to the exchange of photons is related to the charge *e* (see section 8.3) so the strength of the inter-quark force is related to the colour charge. However, there are three of these and, because of this, the symmetry of QCD requires the existence of eight types of gluon. Also, unlike photons which carry no electric charge, gluons carry colour charge. It is this multiplicity of colour-charged gluons which leads to the peculiar properties of the attractive inter-quark force. Roughly speaking the force is essentially zero when the quarks are extremely close; this is known as *asymptotic freedom*. It behaves rather like an electrical force, but around 100 times stronger, when quarks are at their typical separations (around 10^{-15} m) in an elementary particle. Then, for larger separations it behaves like an elastic force becoming increasingly stronger as the separation increases.

It is interesting to see what happens when attempts are made to knock quarks out of elementary particles by bombarding them with other high-energy particles from an accelerator. Suppose the bombarded particle is a meson (consisting of a quark and an antiquark). The energy given to the component quark means that it is pushed apart from the antiquark, the bombarding energy being converted into potential energy in the 'stretched' interquark force. For sufficiently high bombarding energy, this potential energy can then be used to create a quark and antiquark, pair (cf pair production of electron and positron—section 7.6). The system now collapses into two quarks and two antiquarks, which then, because of the attractive inter-quark force, join together to form *two* mesons. At higher energies, even more mesons would be formed and the attempt to obtain an isolated quark continually fails. Similar outcomes result if baryons rather than mesons are bombarded. Gluons also cannot exist alone and the best that can be achieved is to produce, for example, transitory *glueballs* consisting of several gluons bound together for a short period.

The quark theory (QCD) of the strong interaction is now well tested and established and leads to a very full understanding of the properties of baryons and mesons. None the less, much work remains to be done.

9.4 The Weak Interaction

The discussion so far in this chapter has been essentially about the structure and behaviour of the many varieties of hadron. Their quark structure is determined by quantum chromodynamics which is the basic theory of the strong interaction. As has been stressed, strong interactions satisfy *all* the conservation laws we have so far encountered and strong interaction processes, exemplified by the production of new elementary particles in accelerator experiments, are very fast and prolific. They typically happen in times of the order 10^{-23} s. Very heavy hadrons, created in such experiments, only live for times of this same order of magnitude. They decay by means of the strong interaction into lighter hadrons, the decay processes always obeying all conservation laws. However, lighter hadrons such as the Λ-, Σ-, π- and K-particles listed in table 9.1 have no lighter particles to which they can decay without violating one or other conservation law and they live for much longer times—around 10^{-10}–10^{-8} s—apart from those decays involving photons (γ) when the mean lives are much shorter. The latter arise from the electromagnetic interaction and the former, much slower processes, are due to the weak interaction.

Weak interaction decay processes manifest themselves in three forms.

Leptonic Decays. *These are processes which only involve leptons such as*

$$\mu^- \to e^- + \bar{\nu}_e + \nu_\mu.$$

This process is rather like the prototype beta-decay process (see below) except that it only involves leptons. Here it should be noted that both types of neutrino are generally assumed to have very small, if not zero, mass.

Semi-leptonic Decays. *These are processes such as beta-decay which involve both hadrons and leptons. Examples are*

$$n \to p + e^- + \bar{\nu}_e \text{ (beta-decay)}$$
$$\pi^- \to \mu^- + \bar{\nu}_\mu$$
$$\Lambda^0 \to p + e^- + \bar{\nu}_e.$$

Strangeness is conserved in the first two processes but not in the third. The weak interaction allows both strangeness-conserving and strangeness-non-conserving semi-leptonic processes to take place. The semi-leptonic process

$$p + e^- \rightarrow n + \nu_e$$

can also take place and is important in understanding the formation of neutron stars (see section 10.5).

Non-leptonic Decays. *These are processes not involving leptons such as*

$$\Lambda^0 \rightarrow p + \pi^-$$
$$\Xi^- \rightarrow \Lambda^0 + \pi^-.$$

In all such processes the law of conservation of strangeness is violated and the strangeness always changes by ±1.

The foregoing examples are just a few from the very many hundreds of observed weak decay processes in the three categories. They all involve the violation of one or more of the conservation laws which govern strong interaction processes and here we come to another very important violation namely the *non-conservation of parity* (mentioned in section 8.7).

Until 1956 physicists had always assumed that the nature of the physical world was such that the mirror image of any physical process was also allowed in the real world. Watch the everyday world through a mirror and everything that is seen to take place could happen in the real world. Of course, there would be 'give-aways'—lettering would be wrong for example. But our alphabet is only a matter of convention and the fact that it looks different through a mirror is nothing to do with the underlying laws of physics. Similarly our left hand observed through a mirror appears as a right hand but the mirror image person still obeys the basic laws of physics. It had always been assumed that there was no fundamental physical process or interaction which was *intrinsically* left or right handed. If there had been then its mirror image would have had the opposite handedness and would *not* therefore ever be seen in the real world. In other words the laws of physics were assumed to be *invariant under mirror reflection.*

As we have seen in section 9.2 invariance under a process always leads to something being conserved. In this case the 'something' is referred to as *parity* which can either be *even* (+) or *odd* (−). Roughly speaking something with odd parity changes sign when studied through a mirror (a left-handed screw becomes a right-handed screw) whilst something with even parity remains the same (for example, a dumb-bell). All elementary particles have a definite parity and it follows that, if parity is conserved, all particle processes are such that the *product* (parity is not an *additive* conservation law) of the parities at the end of a process is the same as at the outset.

In the 1950s a difficulty had arisen in understanding the K^+ meson which was observed to decay sometimes into two pions and sometimes into three. Pions have odd parity and therefore two together have even parity ($-1 \times -1 = +1$) whilst three have odd parity ($-1 \times -1 \times -1 = -1$) so that since the K^+ has odd parity the two-pion decay does not conserve parity. T D Lee and C N Yang (1957 Nobel Laureates) suggested that the problem could be resolved if the weak interaction in this process, and in general, did not conserve parity. This meant that weak interaction processes would exhibit a 'handedness' and it was suggested that such handedness might be observed in the well-known beta-decay process. In 1957 an experimental study of the way in which electrons are emitted from spinning radioactive nuclei (cobalt-60) was undertaken by C S Wu and her collaborators in the USA. She cooled an assembly of cobalt nuclei down to around 0.01 K so that they were hardly affected by thermal vibrations and their spins could then be aligned to point in the same direction by a powerful magnetic field (see figure 9.2(a)). It was found that the electrons were emitted mainly in the downwards direction, implying that the whole process was left-handed in nature as illustrated. In other words, if curled fingers indicate the direction of spin and the thumb the preferred direction of electron emission, then the left hand must be used to agree with experiment. Further experiments also established that neutrinos in beta-decay were *always* emitted spinning in a left-handed way about their direction of motion like a rifle bullet (see figure 9.2(b)). Right-handed neutrinos are never observed although, as might be expected, antineutrinos *are* right handed. This handedness or, in turn, non-conservation of parity is now well established to be characteristic of all weak interaction processes, but strong and electromagnetic processes all conserve parity and any small non-conservation observed in them can always be attributed to slight modifications of the process by the weak interaction.

In this context it is interesting to note that in 1964 another related invariance principle was found to be violated in the decay of the K_L^0. The K_L^0 and the K_S^0 are listed in table 9.1 and are different combinations of the K^0 and its antiparticle ($\overline{K}^0$). This mixing arises because the weak interaction allows the K^0 and the $\overline{K}^0$ to convert backwards and forwards into one another. The K_S^0 always decays into two pions whilst the K_L^0 usually decays into three pions but just occasionally into two. Detailed theoretical analysis shows that this latter behaviour implies that there must exist a very weak interaction which is not invariant under the reversal of time. Up to this date it had been assumed that the basic laws of elementary particle physics were all invariant under time reversal. Thus, if a video was made of any elementary particle process and it was run backwards, then the observed time-reversed process would also be an allowable elementary particle process. Here it must be

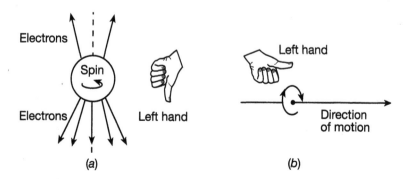

Figure 9.2: *Parity non-conservation in beta-decay: (a) electron distribution; (b) neutrino spin.*

stressed that this invariance does not hold for complex physical phenomena such as steam pouring out of a kettle—the time-reversed process of steam rushing back into the spout would not be seen in the real world. This latter irreversibility relates to the statistical nature of complex systems and the second law of thermodynamics (see section 3.4) and not from time reversal non-invariance of the interactions between basic particles. Much experimental work has been targeted at finding other examples of time reversal non-invariance. In particular it is predicted that the neutron should behave as though it had a small amount of positive electric charge at one end of its spin and an equal amount of negative charge at the other end rather like the north and south poles of a magnet. The predicted effect is, however, very small and so far (1996) has not been detected.

The weak interaction is clearly complicated, not only because of the way in which it violates various conservation laws but also because of the many different types and combinations of particle involved in its manifestation. The next task is to describe how it has now become well understood in terms of an impressive fundamental theory.

9.5 The Electroweak Interaction and Unification

We have seen that the electromagnetic and strong interactions are propagated between interacting particles by the exchange of gauge bosons—photons and gluons respectively—between the particles. It is therefore natural to postulate that the weak interaction is propagated by additional gauge bosons. Since it is experienced by both hadrons and leptons it is to be expected that these gauge bosons will interact directly with the quark ingredients of hadrons and with leptons. Thus the beta-decay of a neutron (quark content udd—see section 9.3) into a proton (quark content uud), involving simply the transformation of a d-quark into a u-quark with the emission of an electron and an antineutrino, is now understood in terms of the process shown in figure 9.3(a). The interaction is propagated by a new gauge boson with negative charge known as the *W-boson*. Similarly the leptonic decay of the muon (see table 9.1) can be understood in

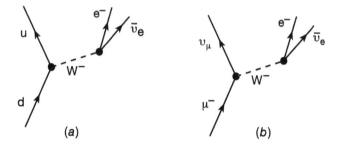

Figure 9.3: *(a) Beta-decay of a d-quark; (b) muon decay.*

terms of figure 9.3(b). The more complex decays inolving only hadrons can be similarly described in terms of the exchange of W-bosons between their component quarks.

At this stage the existence of the W-bosons (they are required to exist with both positive and negative charge) is simply a postulate. However, a remarkable step forward in our understanding of the weak interaction was initiated in 1960 by Glashow, Weinberg and Salam (Nobel Laureates, 1979). It turns out that the weak interaction is not only similar to the electromagnetic interaction through their exchange of gauge bosons—the W-boson and the photon—but is intimately related to it, and that the coupling constant of the W-boson to quarks and leptons is the same as that of photons, namely *e*, the basic electric charge. The 'weakness' of the weak interaction is then accounted for if the W-boson is very heavy—it needs to be about 85 times the mass of a proton. The weakness arises from the immense amount of energy needed to momentarily create the W-boson which, because of the limitations of the Heisenberg exclusion principle, can then only exist for a very short time. The full *electroweak theory* also required the existence of another gauge boson (the Z-boson) with zero electric charge, rather like the photon but very heavy. The existence of this new particle implied that hitherto unobserved weak interaction processes should take place and it was a great triumph for the theory when such processes were observed and when the W- and Z-bosons were actually detected (1983) at CERN in accelerator experiments and found to have the masses predicted by the theory.

This combination of electromagnetic and weak interaction theory, known as *electroweak theory*, is beautiful and its predictions are fully confirmed by experiment. It is highly symmetrical in its formulation and the weak and electromagnetic interactions only exhibit their differences, for example the large difference in mass between the photon and the W- and Z-bosons, at low energies. This is referred to as *spontaneous symmetry breaking*. Spontaneous symmetry breaking occurs in other more familiar physical situations, for example in a bar magnet. At ordinary temperatures the magnetic iron atoms in a magnet line up pointing essentially in the same direction, yet the laws governing this behaviour are perfectly symmetrical and do not pick out any particular direction. The symmetry breaking in a magnet is easily explained in terms of the forces between the atoms which tend to line them up. The electroweak symmetry breaking is similarly attributed to the interaction of elementary particles with a new field pervading the whole of space and which is believed to be responsible for their masses. A theory of such a field was first put forward by Higgs in 1964 and just as the photon is a manifestation of the electromagnetic field so a manifestation of the Higgs field is expected—the *Higgs boson*. It is predicted to be very heavy and so far has not been detected. One of the main incentives for building higher-energy accelerators such as the LHC (see section 9.1) is to produce and detect this particle.

Of course it would be very satisfying if this 'bringing together' of the weak and electromagnetic interactions could be extended to include the strong interaction which has a similar mathematical structure and also involves six basic quarks (see section 9.3) analogous to the six basic leptons (see table 9.1). Again, at the energies available in accelerators today, there are fundamental differences between the three types of interaction. Not least, the strong interaction is some 100 times stronger than the electromagnetic interaction. However we saw in section 9.3 that the inter-quark force becomes weaker when the quarks are very close together. This only happens at very high energies and to reach the situation when all three interactions have the same strength it is predicted that an energy—the unification energy—of the order of 10^{15}–10^{16} GeV is necessary. This is way beyond accelerator energies in the foreseeable future but such energies were around during the 'big bang' as the universe came into being (see section 10.3). This retrieval of the overall symmetry at high energy is rather like the retrieval of full spatial symmetry with a magnet when its temperature is raised to around 1050 K: at this temperature the magnetism, and of course its directionality, disappears. Below the unification energy *spontaneous symmetry breaking* ensues and the three interactions revert to the strengths we observe in the everyday world.

During the last 20 years a major thrust of theoretical physics has been to try and develop complete and satisfactory unified theories of the weak, electromagnetic and strong interactions—so-called *grand unification theories (GUTs)*. Such theories would mean that above the unification energy quarks and leptons would have essentially the same properties and could change one into the other. In turn, this would mean that even a low-energy everyday proton, for example, would have a very small but finite probability of converting into a positron and a neutral pion. This process would violate the laws of lepton and baryon conservation (see section 9.2) and its lifetime is estimated to be at least 10^{30} years. So far it has not been observed. It is also predicted that *magnetic monopoles*—loose north and south magnetic poles—should exist.

One very attractive and popular GUT is based on what is called *supersymmetry* which proposes that basic spin-$\frac{1}{2}$ particles such as quarks and electrons have partners with integer spins and that, correspondingly, gauge bosons (photon, gluon, W, Z) have partners with half-integer spins. A further very promising development incorporating such ideas is to represent the basic particles, which one normally thinks of as 'pointlike' entities, by different vibrations of fundamental one-dimensional strings (referred to as *superstrings*) which may have ends or be in loops—*superstring theory*. Such a theory can only be formulated if space-time has many more dimensions—possibly ten—than the four (three space, one time) we have considered so far. The extra dimensions would be 'curled up' into a very small size and would, therefore be unobservable. Here it is interesting to note that superstring theories also naturally include gauge bosons having spin 2 (cf the graviton—Section 7.7) and thus open the door to the possibility of a unified theory including the gravitational interaction. The possibility of finding such an all-embracing unified theory will be discussed briefly in section 11.2.

9.6 Moving Forward

In this chapter the present-day perception of the nature and interactions of the fundamental constituents of matter have been outlined. It is remarkable that the complexities of everyday physical (and biological) processes can, in principle if not (yet?) in practice, be understood in terms of a few basic particles—quarks, leptons and gauge bosons—and their interactions. In the next chapter this development of understanding of the very large in terms of the small will be taken further as we explore the nature and behaviour of the universe as a whole.

CHAPTER 10

ASTROPHYSICS AND COSMOLOGY

Stars, Galaxies and the Expanding Universe

10.1 An Outline of the 'Visible' Universe

The physics we have dealt with so far has been largely explored and appropriate theories developed using controlled laboratory experiments. These experiments have been used over the years to provide information about a wide variety of physical processes from the sub-atomic through to the macroscopic behaviour of complex materials. In turn, many experiments have been specifically designed to test the predictions of new physical theories and, as a result, revolutionary changes have taken place—for example, the development of quantum and relativistic mechanics. However, in approaching the structure and behaviour of the universe itself, it is no longer possible to 'set up' controlled laboratory experiments. The evolution of the universe and the behaviour of its components—stars, galaxies etc—is a natural ongoing physical process over which we have no control. The best that can be done is to design experiments to study particular aspects of these processes.

Because of the massive size of the universe and the extreme conditions that exist in stars such study can only be conducted by observing physical processes from a distance. The oldest way of doing this is simply by studying the light emitted from stars through optical telescopes which, over the years, have become increasingly large and sophisticated. The largest are *reflecting telescopes* using concave mirrors with diameters up to around 10 m which are placed at high altitudes so as to minimize the distorting effects of the earth's atmosphere. However light is only a minute part of the electromagnetic spectrum (see table 4.1) and celestial objects emit electromagnetic radiation throughout this spectrum. Such radiation is not 'visible' in the strictest sense of the word but is, of course, detectable by appropriate instrumentation. Most parts of this wider spectrum are unable to penetrate the earth's atmosphere but wavelengths in the millimetre to a few metres range are readily detectable and *radio telescopes* using one very large or several smaller discs (like large domestic satellite dishes) have provided much useful information. Radiation in the infrared, microwave and x-ray region has to be detected using instruments carried in rockets or satellites above the earth's atmosphere. Altogether, these different approaches have provided a wide spread of information about the electromagnetic radiation in space and, in turn, about its origins.

In addition, there are other forms of radiation which can be detected. Cosmic rays, which consist mostly of protons together with a few alpha-particles and light nuclei, some of which can have very high energies, are continually bombarding the earth. They probably originate in supernovae and quasars (see section 10.5) and lead to the creation of showers of other particles when they collide with atomic nuclei in the atmosphere. There is also a steady stream of neutrinos from the nuclear reactions in stars (see section 10.4), which pass through the earth and which hardly interact at all because they only experience the weak interaction. In consequence, they are extremely difficult to detect (see section 8.6).

Resulting from a study of the electromagnetic radiation detected in space the following outline description of the nature of the physical universe emerges. We live on the earth, a nearly spherical planet with diameter 1.27×10^4 km, which moves around the sun during the year in an elliptical path at an average distance of 1.5×10^8 km. There are eight other planets of various sizes all orbiting in the same direction and in roughly the same plane—the *ecliptic*—some nearer to (Mercury, Venus) and some further from (Mars, Jupiter, Saturn, Uranus, Neptune, Pluto) the sun than the earth. In addition there is an assortment of debris—*asteroids* (very small planets mostly between Mars and Jupiter)*, meteoroids* and *comets*—also in orbit. The sun is a medium-size star having a diameter of about 1.4×10^6 km and is just one of about 10^{11} stars which are gathered together in the *galaxy* which we see at night as the Milky Way. This galaxy is in the form of a disc with a diameter around 10^{18} km (10^5 light years, where 1 light year is the distance travelled by light in one year, namely 9.46×10^{12} km).

The detailed shape of the disc is unclear but it has roughly the form shown schematically in figure 10.1(a) and is surrounded by a halo of stars. It bulges in the centre and the solar system is about two-thirds of the way to the edge. Our nightly view of the Milky Way is 'edge on' and arises from looking inwards towards the centre of the disc, this view being somewhat obscured by the presence of interstellar dust. The component stars of the galaxy are slowly orbiting about its centre taking around 10^8 years to complete an orbit. Our galaxy is called a *spiral galaxy* since, away from the centre the stars are in spiral arms as shown in figure 10.1(b). There are many galaxies like this, but there are many others which are not in the form of spiral discs and which are ellipsoidal in shape, and then others fall in between these two shapes. They also vary in size, some containing as few as 10^6 stars and others as many as 10^{13}.

There are around 10^{11} galaxies in the visible universe. Many of them are gathered together in clusters, which may contain several thousand galaxies held together by their mutual gravitational attraction. The distribution of galaxies and clusters is rather like that of the material in a sponge: there are sheets and filaments of galaxies and there are also many large gaps. So, although uneven on the small scale, the visible universe appears to be essentially uniform, as with a sponge, on the large scale. This uniformity is referred to as the *cosmological principle*.

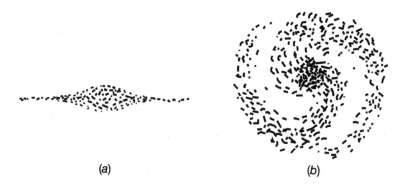

(a)   (b)

Figure 10.1: *A schematic representation of our galaxy: (a) edge on; (b) a full view.*

The foregoing brief description of the visible universe is extremely superficial but it does give a framework within which more detailed discussion of its different features can be conducted. The first task is to consider the information that can be derived from the study of the electromagnetic radiation generated within the universe.

10.2 Electromagnetic Radiation in the Universe

One important point to remember in discussing electromagnetic radiation is that the spread in its frequency is characteristic of the temperature of the emitting source. This was discussed in section 5.2 and, in particular, it was noted that for a body at absolute temperature T there is a spread in the frequency of the emitted radiation which peaks at a frequency around $10^{11}T$ oscillations per second. The visible optical radiation lies in the frequency range 4×10^{14} (red) to 7×10^{14} (blue) (see table 4.1), implying that the surface temperature of a star emitting such radiation must be in a range somewhat greater than 4000 K to 7000 K: roughly 3000 K–10,000 K. Because of the spread in frequencies emitted, the colour of a star is not precisely that given by the peak frequency above. For example, a star at 3000 K will indeed be reddish but one, like the sun, at around 6000 K will emit all the colours of the rainbow to some degree, which then combine together to give white light (see section 4.5) as observed. In discussing the formation and nature of stars in section 10.4 we shall come to understand how these large surface temperatures arise.

Turning now to lower frequencies we come to radiation in the infrared and microwave region. The former arises largely from heated dust lying within galaxies, the heating being due to hot radiation from stars in the process of forming or dying. These processes will be discussed in section 10.4.

As far as microwave radiation is concerned, a most important observation was made in 1965 when Penzias and Wilson (Nobel Laureates, 1978) using a very sensitive detector discovered such radiation clearly arriving at the earth from outside our galaxy. They knew the radiation was extra-galactic since its intensity remained the same day and night throughout the year; if it had been coming from the sun or our galaxy the intensity would have changed as the direction of the detector changed with the earth's motion. Subsequent investigation culminating in a detailed study in 1989 using the Cosmic Background Explorer (COBE) satellite established that the whole of space is pervaded with microwave radiation (*cosmic microwave background radiation*) having exactly a distribution of frequencies (see section 5.2) corresponding to a temperature of 2.736 K. The presence of this background microwave radiation has profound implications for understanding the way in which the universe came into being which will be discussed in the next section.

At even lower frequencies we come to the part of the electro-magnetic spectrum in the radio-wave region. From the 1940s onwards many thousands of discrete radio sources have been identified. Many have been found to coincide in position with known galaxies, and radio waves from within our own galaxy have also been detected. All such radiation (synchrotron radiation—see section 9.1) can be understood as due to the acceleration of charged particles, particularly electrons, in the very large magnetic fields which exist within galaxies.

Finally, turning to frequencies higher than those for optical radiation, we come to extreme ultraviolet and x-ray astronomy (see table 4.1). This has developed rapidly with the use of satellites carrying appropriate detectors and, for example, many x-ray sources have been pin-pointed, mostly outside our own galaxy. Given that x-ray frequencies are around 10^{17} oscillations per second upwards (see table 4.1) it follows from the above relation between peak frequency and temperature that the objects emitting x-rays must be at extremely high temperatures—around 10^6 K–10^7 K. They are obviously not ordinary stars, which have much lower temperatures (see above), and in section 10.4 the possible origins of this radiation will be discussed. Even higher-frequency radiation (gamma-radiation—see section 8.6) has also been detected by satellite observations emanating from the central disc of our galaxy, implying temperatures around 10^9 K. This is attributed simply to collisions between very high-energy cosmic ray particles and the dust mentioned earlier which lies between stars.

From this brief survey of the electromagnetic radiation present throughout the universe it is clear that many questions about its origins have to be answered. Much depends on the way in which stars are formed and develop. But before considering this it is important to discuss how the universe as a whole has developed into its present state.

10.3 The Expanding Universe and the Big Bang

Looking out into the universe with an optical telescope and studying in detail the spectrum of light emitted by stars in other galaxies than our own a most remarkable situation emerges. First of all the broad spectrum of light emitted by any star has a number of dark lines in it, implying that light of the corresponding frequency is being absorbed. Such absorption is readily understandable as being due to atoms in the stellar atmosphere and, as discussed in section 5.6 (see figure 5.5(b)), the absorption spectrum should be characteristic of these atoms. Very recognizable patterns were indeed observed in the light from stars in our own galaxy. However, it was found that for distant stars the patterns of lines were not those of any known atoms. This mystery was solved when it was realized that the patterns were indeed identical with those of known atoms *but with the frequencies all proportionally decreased by exactly the same amount.* Since the shift to lower frequencies (and, therefore, longer wavelengths) is towards the red end of the spectrum it is universally known as the *red shift.* Such a shift is readily understandable if the corresponding stars are moving away from us at a very high speed and is caused by the Doppler effect discussed in section 2.6.

An intensive study of the speeds with which stars are receding in relation to their distance from us was made in the 1920s by Edwin Hubble. One way of estimating this distance was to study stars whose light output varies periodically (Cepheid variables). This variation is intimately related to their luminosity—that is, the total amount of light energy they emit—a relationship which was well established for stars in our own galaxy. Hubble identified similar variable stars in distant galaxies which were, however, much dimmer because of their greater distance away. Knowing how much light they were emitting from the above relationship, he was able to estimate their distance from the extent to which the light received had been diminished in travelling to the earth. Further, the magnitude of the red shift they exhibited enabled their recessional speeds to be deduced. He found that this speed was simply proportional to their distance away; the further away a galaxy is, the faster it is moving away from us—typically around $5\times10^7\,\mathrm{m\,s^{-1}}$ for a galaxy at a distance of 3×10^9 light years. This relationship is embodied in what is known as *Hubble's Law* and means that the universe is not static but is in a continual state of expansion. This is not to say that we are at the centre of the universe, but that space–time is curved because of its matter content (see section 7.7) and all parts of the universe are moving away from each other, rather like the way in which spots on the surface of a balloon move apart as the balloon is inflated. The further apart two spots are the faster they move away from each other. The universe thus looks roughly the same overall to an observer from whichever galaxy the observation is made.

Since galaxies are in general rushing away from each other the implication is that in the past they were much closer and that at some point there must have been a gigantic explosion or fireball— the *hot big bang*—setting this expansion of the universe into motion. From Hubble's Law it is possible to estimate how long ago in the past the big bang took place. A simple calculation gives a figure around 20 thousand million years (2×10^{10} years). However, this does not take into account the gravitational attraction between the matter in the universe, which tends to slow down the rate of expansion and, taking this into account and recent observations from the Hubble satellite telescope, a smaller figure is obtained— possibly as low as 9×10^9 years. This age of the universe should be consistent with the age of our own galaxy. This can be estimated, for example, by comparing the relative abundances of radioactive isotopes (see sections 8.1 and 8.6) of uranium in the earth which decay at different rates. This enables the time of their formation to be estimated as being around 1–1.5×10^{10} years ago. If the lower estimate for the age of the universe is correct then there may be a problem since the age of our galaxy appears to be greater than 9×10^9 years. Fortunately, very recent (1997) measurements now suggest an age for the universe of around 1.4×10^{10} years.

There is another important observation which ties in with the big-bang hypothesis, namely, the cosmic background radiation discussed in the previous section, which apparently pervades the universe. At the time of the big bang all the energy now possessed by the universe was concentrated at a point—called a singularity—in the initiating explosion. Space–time initially had zero extension and simply 'grew' with the explosion. Initially the temperature and the associated energy must have been infinite. The energy would clearly be above the grand unification energy (see section 9.4) and the situation would be one unknown in our everyday world. After about 10^{-10} s, however, the temperature would have dropped to a few million billion degrees ($\approx 10^{15}$ K) and the corresponding particle energies would have been around a few 100 GeV. At this energy spontaneous symmetry breaking would have taken place and the usual strong, electromagnetic and weak interactions would be in operation. Associated with such temperatures, and produced in the explosion, would be electromagnetic radiation of extremely high frequencies—in the x-ray and gamma-ray region (see table 4.1). Then, as the universe expanded, the wavelengths of this radiation would be continually stretched by the expansion, leading to increasingly longer wavelengths and, therefore, lower frequencies. In other words, the initial burst of electromagnetic radiation characteristic of a very high temperature cools down and now it is perfectly reasonable that its frequency distribution today, after such a long elapse of time, should be characteristic of a very much lower temperature—the 2.736 K observed.

Finally, before leaving this broad discussion of the expansion of the universe we must briefly consider its possible future. The rate at which the universe continues to expand is conditioned by the amount of matter it contains—the average density of matter—since, as mentioned earlier, the gravitational attraction between pieces of matter, for instance between galaxies moving apart, holds them back and slows down the rate of expansion. The denser the matter, the more powerful is this effect. Basically there are three possible scenarios rather similar to the three possibilities which arise when an object is projected away from the earth. In this latter case, if the speed of projection is slow the speed will decrease to zero and then the gravitational attraction will pull the object back to the earth; this is what happens when a ball is thrown into the air. If the speed is very high, the gravitational attraction is not strong enough to hold the object back and it will escape from the earth and go off to infinity as in the launch of a spaceship. Finally, there is the situation that the speed with which an object is projected has precisely the critical value which *just* enables it to escape and *just* make its way to infinity.

So with the universe (see figure 10.2). If its matter density is high enough, then the expansion will eventually be drawn to a halt and the universe will start collapsing in on itself leading ultimately to the opposite of the big bang, namely the *big crunch*. In general relativistic terms (see section 7.7) space–time is curved back on itself by the gravitating matter and the universe is finite but unbounded, analogous to the two-dimensional surface of a sphere which is also finite but unbounded. If the matter density is low, then the expansion will continue indefinitely and space–time is again curved but becomes infinite in extent. If its value is what is called the *critical density* (on average just a few atoms per cubic metre) then the universe just manages to expand all the way to infinity in an infinite time. Space–time is then effectively 'flat'. A

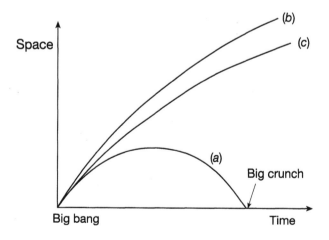

Figure 10.2: *Expansion of the universe for different matter densities: (a) high density; (b) low density; (c) critical density.*

crucial question is therefore what, in fact, is the density of matter in the universe?

Another related issue is the value of the cosmological constant referred to in section 7.7. The observed rate of expansion of the universe implies that this constant is extremely small. However, it can be shown that there is a very large additional contribution to this constant due to quantum fluctuations (related to the Heisenberg uncertainty relation) and this implies that the constant originally inserted by Einstein must be virtually equal, but opposite in sign, to this quantum contribution. This cancellation is estimated to be accurate to 1 part in 10^{22}! Why this amazing cancellation occurs is not yet understood. Neither is it known whether it is exact. There are clearly some fascinating issues concerning the evolution of the universe. Some of them will be addressed in section 10.5 and in Chapter 11 after we have discussed the different forms of matter which occur in the universe and which contribute to this density.

10.4 The Early Stages of the Universe and the Formation of Stars

We have seen in the last section that there is good evidence that the universe, including space and time itself, came into being with the big bang. The temperatures and energies involved were initially immensely high and the universe was an indescribable 'soup' of matter and radiation. After about 10^{-10} s, however, the temperature dropped to around 10^{15} K and the corresponding energy was such that spontaneous symmetry breaking set in and the usual strong, electromagnetic and weak interactions came into operation. However, atoms and nuclei could not have existed—particle energies would have been too high for them to have held together. There would have been simply the most basic of the elementary particles we have encountered. Most prolific would have been light particles such as electrons, positrons, neutrinos and their antiparticles together with photons. These would have been in a state of continual change as they interacted with each other through creation and annihilation processes (see section 7.6). In addition, there would have been a few (roughly 1 part in 10^9) heavier particles, particularly protons and neutrons.

As the universe continued to expand the temperature is estimated to have dropped from 10^{15} K down to 10^{10} K after one second and then, more slowly, down to 10^9 K after a few minutes. At this latter temperature there would not have been sufficient disruptive energy around to prevent the neutrons and protons present joining together under the attraction of the nuclear force to form heavy hydrogen nuclei (deuterons—$^2_1$H) consisting of one proton and one neutron. In turn, these simple nuclei could then quickly combine with further neutrons and protons eventually to form the very stable helium nucleus ($^4_2$He) consisting of two protons and two neutrons. Further, at this point in the evolution of the universe, most of the electrons and positrons would have annihilated each other to form photons and many neutrons would have beta-decayed (see section 8.6) so that the contents of the universe would be mainly photons, neutrinos and antineutrinos together with a relatively small number of protons, helium nuclei and electrons and an even smaller number of deuterons. It is estimated that roughly 25% of the mass created in the first few minutes of the big bang was in the form of helium nuclei and the big-bang theory is well supported by the fact that this percentage is found in present-day studies of the matter distribution in the universe.

After the first few minutes the universe continued to expand rapidly. The cooling process continued so that after a few hundred thousand years its temperature was so low that it was possible for electrons to join up with the protons and helium nuclei to form hydrogen and helium atoms; the photons no longer had sufficient energy to knock the atoms to pieces. Thus, at this stage, matter in the universe was mostly in the form of a cloud of hydrogen (about 75% by mass) and helium (about 25% by mass) gas. Under the influence of their mutual gravitational attraction the gas atoms were then gradually drawn together to form stars and galaxies. Here it is interesting to note that small variations recently observed in the background electromagnetic radiation are indicative of early variations in the density of the universe, which would have acted as focal points for their formation. This falling together means that gravitational potential energy is converted into kinetic energy of the atoms, leading to a continual increase in the temperature at the centre of a star so that atoms are ionized—

the electrons are knocked out—and a plasma of ions and electrons is formed. Eventually the temperature of the core becomes so high, around 1.5×10^7 K, that the kinetic energies of the hydrogen nuclei (protons) are sufficient to overcome the electric repulsive forces between them and to set in train the 'hydrogen burning' fusion process described in section 8.5. The net effect of the basic fusion process is that four protons are converted into helium with the emission of two positrons and two neutrinos and the release of energy equal to about 25 MeV. In the sun, the rate of energy generation from hydrogen burning is around 5×10^{20} MW, a factor of 10^{11} larger than the level of energy production in the USA! This energy is transmitted to the outer layers of the star and emitted as electromagnetic radiation.

Thus the structure of the sun is of the form shown in figure 10.3. It consists of the central core generating an immense amount of energy through nuclear fusion surounded by a large zone consisting of plasma through which energy is transmitted to the surface and in which convection currents continually feed protons

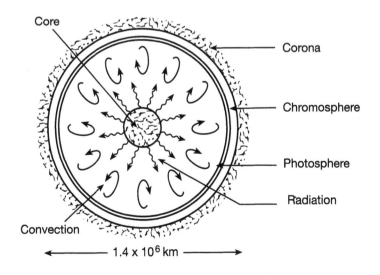

Figure 10.3: *The structure of the sun.*

into the core. The surface, known as the *photosphere*, is at a temperature of around 6000 K leading to the emission of the star's light. Beyond the surface is the *chromosphere*, which is a layer of rarefied gases, and beyond that is the *corona*, consisting of extremely fine dust particles at a very high temperature. The sun is also emitting streams of particles, particularly protons and helium nuclei, into space and is losing mass at the rate of 1 part in 10^{13} or 10^{14} in a year. This emission is known as the *solar wind* and, when it interacts with molecules in the earth's atmosphere under the influence of the earth's magnetic field, leads to such phenomena as the *aurora borealis*—the northern lights—and geomagnetic storms.

10.5 The Lives of Stars

This 'burning' of hydrogen to form helium continues for a long time—around 10^{10} years for a star like the sun, which is currently about halfway through this burning process. Most observable stars are in this state and are referred to as *main sequence stars*. However, after some 10% or so of the star's mass has been converted into helium, the high-temperature core begins to contract and the outer layers of the star undergo massive expansion and cool down. In this state, the star is very large and, correspondingly, it is referred to as a *red giant*. The contraction of the core means that more gravitational energy is released, leading to a further increase in its temperature to around 10^8 K. This enables 'helium burning' to take place, which leads to the formation of very stable carbon ($^{12}_{6}$C) nuclei—these are, as it were, a combination of three helium nuclei. Here it should be noted that the combination of two helium nuclei (beryllium—$^{8}_{4}$Be) is unstable and is only formed transitorily in the build-up of the carbon nuclei. It is also remarkable that the formation of the carbon nuclei only happens at a significant rate because there is an energy level in the carbon nucleus at just the right energy to enable a transitory beryllium nucleus to capture a helium nucleus by a strong resonant process (see section 8.5). This fortuitous energy level is a key factor in the existence of life and will be referred to again in section 11.2.

Many other nuclear reactions in stars can take place, leading to the synthesis of even heavier nuclei such as silicon and iron as further collapse of the core takes place and higher temperatures are reached, but the nature of this collapse and the further evolution of a star depends on its mass and there is a variety of possible scenarios.

Brown Dwarfs. First it should be mentioned that if the mass of the matter gathered together by gravitational attraction is less than about $\frac{1}{15}$ of the mass of the sun then the gravitational energy released in the collapse and the resultant core temperature is too low to sustain a fusion reaction. Such stars are known as brown dwarfs. Because of their low temperatures they hardly radiate and so are difficult to detect; only a few have been identified.

White Dwarfs. If a stars mass is greater than $\frac{1}{15}$ but less than about $\frac{3}{2}$ times the mass of the sun (known as the Chandresekhar (Nobel Laureate, 1983) limit) there comes a point at which the gravitational force causing the collapse of the core is balanced by the effective repulsion between electrons in the core due to the exclusion principle (see section 5.5)—no two electrons can be in exactly the same state. Thus the core stops collapsing and settles down into a final state with a size similar to that of the earth but with a density some 10^6 or 10^7 times greater. Such a star is known as a white dwarf. The external envelopes of these stars are ejected and become what are known as *planetary nebulae*.

Neutron Stars and Pulsars. If the star's mass is between the Chandresekhar limit and about two solar masses then another end state is possible. Because of the larger mass and the resultant increased effect of gravity, the exclusion principle repulsion between the electrons in the core is overcome and they are captured by protons. This process is due to the weak interaction (see section 9.4), and leads to the formation of neutrons and neutrinos. The latter escape and the core is then simply like a very large nucleus consisting basically of neutrons. The exclusion principle repulsion between the neutrons now stops complete collapse. Even so, the final neutron star only has a radius of a few kilometres and a density some 10^{15} or more times that of the earth—hundreds of millions of tons per cubic inch! Many neutron stars are rotating rapidly and, as a result of this and their magnetic fields, emit regular pulses of radio waves like a lighthouse. They are referred to as *pulsars*. The first to be discovered, in 1967, rotated once a second but many more have now been identified rotating with frequencies up to 1000 times a second.

Black Holes. For even larger masses the repulsion between the neutrons due to the exclusion principle loses out to the gravitational force and further collapse of the core takes place. The gravitational field at the surface is now so strong that light signals cannot escape (refer to the discussion of the effect of a gravitational field on light in section 7.7). The light can only go so far—to the *event horizon*—and is then pulled back like a ball being pulled back to the earth. Anything trespassing within the event horizon can never escape. Since no light emerges from it, the core is referred to as a *black hole* and, in general relativity language, space–time in its vicinity is curved back on itself and emitted light follows this curve. The gravitating matter in the black hole is concentrated into a point of infinite density—a *singularity*—and the only information we can have about it is its mass, angular momentum and electric charge. Since black holes cannot be observed directly their presence in the universe has been deduced from the effects of their huge gravitational fields which extend beyond the event horizon. For example, x-rays are produced as matter is dragged into a black hole in binary star systems (pairs of stars held together by their gravitational attraction) in which a black hole is one (invisible) component. Supermassive black holes having masses of up to a billion solar masses are almost certainly the explanation of *quasars* (quasi-stellar objects). These are extremely bright and distant objects at the centre of some galaxies, including, possibly, our own. They are sometimes 1000 times the brightness of a whole galaxy, and this brightness can be attributed to the radiation emitted as galactic material is continually sucked into such a black hole at the galactic centre. The whole of the central region of the galaxy is presumed to be in a continual state of gravitational collapse into the black hole.

It should also be mentioned that primordial black holes of mass less than that of the sun and which would require gigantic compression of matter may have been created during the initial stages of the big bang but there is, as yet, no concrete evidence for them. Finally, although black holes suck everything into them there is a mechanism, suggested by Hawking in 1973, whereby they can lose energy. In section 7.6 it was pointed out that particle–antiparticle pairs are continually appearing and disappearing in space. When this happens near a black hole the particle could, for example, be sucked into the hole by its gravitational field whilst the antiparticle escapes, taking energy from the black hole with it, thus reducing its mass. So, gradually, all black holes evaporate!

Supernovae. Although the shedding of the outer regions of a star as it becomes a white dwarf is relatively peaceful this is not necessarily so with stars whose mass is greater than the Chandresekhar limit. When the core of such a star collapses into a neutron star or a black hole the fusion process for creation of energy has come to an end and the outward pressure on the external regions of the star decreases so that these regions collapse in towards the core. Their temperature then increases massively and, since they still contain combustible material, nuclear fusion again sets in. The resultant release of energy—up to 10^9 times the previous output—leads to a catastrophic explosion in which the outer layers are blown into space carrying with them many of the elements synthesized during the fusion processes that have taken place and in subsequent neutron-induced reactions. Some of this debris, consisting of both light and heavy elements, will be absorbed as second- and third-generation stars are formed from the interstellar gas and dust. The sun and its planetary system were formed in this way and, indeed, but for this debris the earth (and we!) would not be here. There is also a huge release of electromagnetic radiation and a *supernova* can be brighter than the whole of its galaxy for a few weeks. Many radio sources have been identified as supernovae but only a few optical sources. Here the most famous is the supernova observed in 1034 whose remnants are what we now know as the Crab nebula.

10.6 Problems and Conjectures

In the foregoing sections a desciption has been given of what might be called the 'standard theory' of the evolution of the universe. The idea of the big bang is supported experimentally by the observed expansion of the universe, the all-pervading presence of microwave background radiation and the relative abundance of helium and hydrogen. However it is not all plain sailing and there are several important issues which require further discussion.

First, there is the remarkable uniformity of the background radiation (section 10.2) intensity which, at any point in the universe, is the same to better than one part in a thousand. In section 10.4 we saw that until a hundred thousand years or more had elapsed after the big bang there were free electrons in the universe. These would have interacted strongly with electromagnetic radiation and it was not until they had been absorbed into atoms that the radiation would have been 'free' as we now see it, but when this occurred different parts of the universe were 'out of touch' with each other since they were so far apart that there was not time for communication, even at the speed of light, to take place. So it is hard to understand how the uniformity arose since there could have been different developments in these 'out of touch' regions. A solution to this problem, originally proposed by Guth in 1980 and subsequently much developed and discussed, is that, for a very brief period, the very early universe roughly doubled its size every 10^{-34} s. This is referred to as the *inflationary universe*. The nature of this sort of rapid expansion is illustrated in figure 10.4 and is very different from the rate of expansion for the early stages of the universe illustrated in figure 10.2. Such an inflationary rate of

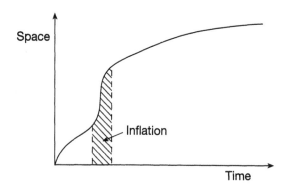

Figure 10.4: *Inflationary expansion of the universe.*

expansion is absolutely formidable and would correspond, for example, to something the size of a tennis ball expanding to the size of the observable universe in only 10^{-32} s! In this time it would have successively doubled its size one hundred times. This means that the universe could have been very much smaller in its early stages so that the different regions would no longer have been 'out of touch' and the uniformity of the background radiation can, in these terms, be understood.

Second, we have seen (section 10.3) that the future of the universe depends on its energy density and that there is a critical value for which space–time in the universe is 'flat'—the expansion, as it were, continually slows down as the universe heads towards infinite size. Estimates of the amount of matter in the universe including that which is shining (stars etc) and that which is so-called *dark matter* (e.g. dust, brown stars, additional matter deduced to be present from its gravitational effect on the motion of stars in galaxies) suggest that the energy density is somewhat less than the critical value. However, it may be that masses of neutrinos are not exactly zero (see section 9.4) and the vast number of neutrinos pervading space following the big bang could then bring the energy density of the universe up to the critical value. Here it is interesting to note that the number of electron neutrinos (v_e) detected on earth emanating from the fusion processes taking place in the sun is found to be less than expected. This could be due to them having a finite mass which would allow some of them to change into muon neutrinos (v_μ—see section 9.4) on their way to the earth and so be undetected. It might also be that additional particles, which would be necessary to achieve unification of the weak, electric and strong interactions (see section 9.5), are around in the universe, contributing additionally to its energy density. What is remarkable is that although the energy density could, in principle be millions of times bigger or smaller than the critical value, the evidence is that it is fairly close to it. It is therefore attractive to feel that it might have *precisely* that value. Here it is interesting to note that the inflationary process just discussed can, in fact, lead to the universe having precisely the critical energy density. The attraction of this possibility is further increased by the fact that in such a universe the (positive) energy due to the mass in the universe (mc^2) is exactly balanced by the (negative) energy due to gravitational attraction so that the net energy of the universe is zero!

Third, we saw in section 10.4 that in the earliest stages of the big bang the number of photons outnumbered the number of protons by a factor of the order of 10^9. These photons would have had extremely high energy and could therefore have produced endless pairs of protons and antiprotons through the pair creation process mentioned in section 7.6 in connection with the production of electrons and positron (antielectron) pairs. Thus the natural expectation would be that the universe should contain equal amounts of matter and antimatter, which in due course would annihilate each other, rather than being entirely composed of matter as seems to be the case. It is encouraging that the idea of unification discussed at the end of section 9.5 with its possibility of baryon and lepton non-conservation, coupled with the time reversal non-invariance properties of the K^0 meson (see section 9.4), may provide a way of creating a slight imbalance between matter and antimatter so that, after mutual annihilation, there is only the residual matter of our present-day universe. Conversely, the existence of a 'matter only' universe can be taken as supportive evidence for the form of unified theories being suggested.

10.7 Moving Forward

So, the big-bang theory of the universe is in a reasonably healthy state. There is much supportive observational evidence but there are also many conjectures still to be refined and tested. Clearly the theory brings together in a remarkable way the nature of the smallest-scale phenomena we know about—the elementary particle interactions—and the large-scale functioning of the universe. Is it possible that there is, indeed, a 'theory of everything' encompassing both aspects? That and other more nebulous and general topics is our concern in the next chapter.

CHAPTER 11

REFLECTIONS ON PHYSICS AND PHYSICISTS

What has Physics Achieved?

11.1 Gathering the Threads Together

Having dealt with the main features of physics at all levels of scale we now go back to some of the general issues first raised in Chapter 1, but first it is helpful to summarize very briefly the areas of physics which have been covered in the previous chapters.

In Chapters 2–6 we saw how, in principle, the mechanical, thermal and electromagnetic properties of inanimate physical matter can be understood in terms of a basic perspective which describes such matter, indeed *all* matter, as an assembly of interacting atoms and molecules. The nature of the interaction which accounts for these properties was understood in terms of the 'electrons plus nucleus' structure of individual atoms and the consequent interatomic force. This force and the atomic structure itself were accounted for in terms of the electromagnetic interaction between the electrons and the nuclei and the quantum theory of Schrödinger and Heisenberg. In addition, taking account of the weak gravitational force, the behaviour of large-scale matter such as planetary motion could also be understood. Overall it is a relatively simple theoretical description which satisfactorily explains a great deal of the physical world around us—the nature and properties of materials and electromagnetic radiation (e.g. visible light, x-rays, radio and TV waves) which are all part of our everyday life. Among other things it is characterized by a number of 'conservation laws'—conservation of energy, momentum, electric charge and so on.

Even this far, however, the introduction of quantum theory, which is essential to achieve understanding, brings with it a feature alien to 'common sense', namely that it is impossible to be absolutely precise about everything. The certainty of classical mechanics had to give way to the uncertainty of quantum mechanics and no longer, for example, can it be specified *precisely* where something is and *precisely* what it is doing. This is not a weakness in experimental method but a fundamental uncertainty intrinsic to the basic theory of quantum mechanics. It relates intimately to the realization that particles can in some sense behave like waves and waves like particles (e.g. photons)—so-called wave–particle duality.

Parallel to this quantum shake-up in classical mechanics was the further realization that classical mechanics also fell down when dealing with speeds near that of light and that it had to be replaced by Einstein's relativistic mechanics (Chapter 7). This brought with it the inevitable, and contrary to common sense, conclusion that different observers in uniform relative motion have different perspectives not only on spatial position but also on the passage of time—time can no longer be regarded as absolute. Further, there emerged a new relationship between mass and energy ($E = mc^2$) and the replacement of the law of conservation of energy by the law of conservation of mass–energy. Since quantum mechanics was based on the former law it was natural to develop a relativistic version and, in achieving this, Dirac predicted that electrons (and for that matter other particles not then identified) should have associated with them antiparticles having the same mass and spin but with every other property (e.g. charge) having the opposite value. A final development at this stage was the generalization of relativity theory—general relativity—to take on board the behaviour of space and time when phenomena were measured in frames of reference no longer in uniform motion relative to each other. In turn this led to a description of space–time which becomes 'warped' in the presence of a gravitational field.

The next step forward (Chapter 8) was the realization that atomic nuclei themselves had a structure and consisted of neutrons and protons held together by a very powerful nuclear force. Just as the electromagnetic force can be understood in terms of the exchange of photons between particles, understanding of the nuclear force was achieved by attributing it to a strong interaction brought about by the exchange of virtual particles (e.g. pions) between the neutrons and/or protons. The transitory existence of these 'messenger' particles and the associated energy to create them (via $E = mc^2$) was allowed by virtue of the short-term uncertainty in energy permitted by the quantum mechanical uncertainty just referred to. Different nuclei and their properties could then be understood in terms of the quantum mechanical behaviour of different combinations of protons and neutrons interacting through the nuclear force. Although it was, and is, difficult to calculate the very fine details of nuclear behaviour in these terms there is no doubt that a deep understanding of nuclear properties—energy levels, nuclear reactions, fission, fusion and radioactivity—has been achieved. In understanding nuclear beta-decay, however, a new interaction had to be introduced, now referred to as the weak interaction, which led, for example, to the decay of a neutron into a proton, and electron and a new (possibly massless) particle called an antineutrino. Thus began the realization that there are basic particles other than the electron, neutron and proton and that apart from the gravitational and electromagnetic interactions there is also a strong and a weak interaction operative in the physical world.

The discovery of the wide variety of new elementary particles (Chapter 9) escalated rapidly as particle accelerators of higher and higher energy were developed. These particles were characterized by their mass, charge, spin and the way in which they experienced the strong, electromagnetic and weak interactions. Literally hundreds have been identified and it was a great step forward when it was found that the complicated and wide spectrum of strongly interacting particles (baryons and mesons) could be simply understood as being different combinations of only a few even more fundamental entities—the six quarks and their antiparticles. Further, just as the electromagnetic force results from the exchange of photons, so the inter-quark force was explained in terms of the exchange of gluons. Similarly, the weak interaction involves the exchange of W- and Z-bosons. It was also found that in addition to the electron and its neutrino, which do not experience the strong interaction, there were four other 'leptons'—the μ and τ and their companion neutrinos. Thus the tangible entities from which the physical world in all its complexity and variety is consititued seem to have boiled down to essentially six quarks, six leptons and a number of gauge bosons—gluons, photons and W- and Z-bosons—propagating the strong, electromagnetic and weak interactions respectively.

Finally (Chapter 10), in parallel to these developments in understanding the fundamental components from which the physical world is constituted, was the understanding being achieved of the behaviour of matter on the large scale—the structure of stars and galaxies—and the evolution of the universe as a whole. As far as the latter is concerned the evidence is very clear that it is in a state of continual expansion and this points inevitably to the postulate that it started this expansion from an unimaginable singularity referred to as the 'big bang'. Whether this expansion will continue indefinitely or eventually reverse itself and finish up as a 'big crunch' is a matter still under discussion.

11.2 Theories of Everything

There is no doubt that full understanding of the physical universe, if it is to be achieved, will depend on the bringing together—unification—of the theories of the behaviour of the basic elementary particles from which the universe is constituted and the properties of space–time and gravity as embodied in the theory of general relativity. The time, energy and length characteristic of such a unification can be estimated using the three key constants of nature (see the Glossary) which are involved, namely $\hbar$ (Planck's quantum constant divided by 2π), c (the speed of light) and G (the gravitational constant). By combining these constants in ways to produce quantities having the appropriate dimensions it is possible to obtain what are known as the Planck time, the Planck energy and the Planck length. They are given by

$$\text{Planck time} = \sqrt{\hbar G/c^5} \approx 10^{-43}\,\text{s}$$

$$\text{Planck energy} = \sqrt{\hbar c^5/G} \approx 10^{19}\,\text{GeV}$$

$$\text{Planck length} = \sqrt{\hbar G/c^3} \approx 10^{-35}\,\text{m}.$$

Such an extremely small time and length and such an extremely large energy do not feature in laboratory experiments—they are quite unattainable. But they would certainly have been experienced in the big bang as the universe came into being. At times shorter than 10^{-43} s the universe had a size less than around 10^{-35} m. The energy was so high (10^{19} GeV or more) that all four interactions—strong, electromagnetic, weak and gravitational—were indeed unified. However, over a period equal to the Planck time, the Heisenberg uncertainty relation (see section 5.6) introduces an energy uncertainty ($\hbar$ divided by the Planck time—see page 83) equal to the Planck energy. This is commensurate with the total energy of the universe at that time and it therefore becomes difficult to discuss the situation in formal theoretical terms. At later times the uncertainty reduces, spontaneous symmetry breaking sets in and, eventually (after about 10^{-10} s—see section 10.4), we are left with the interactions we are used to and a universe consisting of basic elementary particles and electromagnetic radiation (see section 10.4). The universe as we know it gradually comes into being. Terrestial studies of elementary particle processes can at present only be carried out up to energies of a few hundred GeV volts and it is inconceivable that machines producing energies up to the Planck energy will ever be devised. So the study of the early stages of the universe is the most likely source of further information about the behaviour of elementary particles at such very high energies and, in turn, of complete unification.

Whether an ultimate 'theory of everything' embracing gravity and grand unified theories (GUTs) of the three elementary particle interactions will eventually emerge is a matter for conjecture. Certainly there have been promising developments over the last 20 years using supersymmetric string (superstring) theories. As mentioned in section 9.5 such theories are attractive since they naturally include spin-2 gauge bosons and so can accommodate gravity. This accommodation actually requires the strings to have a size around 10^{-35} m. Such theories also have the virtue that the infinities which plague the mathematics of point particle theories seem to be mostly eradicated. In addition particles having a definite handedness as observed (see section 9.4) are also pre-dicted. Because of these and other factors some form of superstring theory is generally regarded as the most likely vehicle for formulating an all-embracing theory of everything. Such theories are currently based on the assumption that quarks are basic components. However, there have been slight indications recently that maybe they themselves have some form of structure. If this turns out to be so then a great deal of rethinking will be required. Such uncertainties underline the point made in section 1.2 that all theories, however much confidence there is in them, have to be regarded as provisional.

11.3 The Anthropic Principle

Beyond this, however, there are many other unanswered questions. Was there only one big bang or were there many and are there, in turn, many universes? Or, if there was only one big bang, did it lead to the creation of many universes? If there are many universes is communication possible between them? Are there 'wormholes' acting as short cuts between parts of our universe which are very distant in space and time or, indeed, between different universes if they exist? Do wormholes, if they exist, open the door to the possibility of time travel? If there are many universes do the constants of nature (e.g. G, h, c, e, . . .) have the same values in all of them?

In this latter connection it should be noted how remarkable it is that the laws of nature and the values of their associated constants are just right to ensure the existence of intelligent life. A most famous example is that, if the constants were slightly different, the energy level in carbon, which is crucial to the formation of carbon through helium burning in stars (see section 10.5), would have a different value and it is unlikely that carbon would then be formed. Yet carbon is a key element in all life forms and its existence is vital. However this is only one of many examples of apparently necessary conditions for the existence of life which would not be met in a slightly different universe. Here might be mentioned the following:

- the lifetime of stars and planets is long enough to allow the evolution of intelligent life;

- neutrons are slightly heavier than protons and convert into them via the weak interaction, so ensuring a plentiful supply of hydrogen (H) and, in turn, water (H_2O);

- the remarkable value of the strength of the nuclear force, which is just strong enough to hold the deuteron (neutron plus proton) together, so enabling the formation of helium through fusion, but not strong enough to hold two protons together—if the latter had not been so, then few protons would have been left over a few minutes after the big bang;

- the magnitudes of the (opposite) electric charge (e) of the proton and the electron are the same to an accuracy of around 1 part in 10^{21} so that the universe is essentially electrically neutral. A very small difference in these charges would lead to the existence of powerful repulsive forces in the universe much stronger than the attractive force due to gravity;

- the effective cosmological constant (see section 10.3) is essentially zero.

These few examples, and there are many more, indicate how delicately geared our universe is to ensure the evolution of intelligent life, so much so that this gearing has been enshrined in what is known as the *anthropic principle*. This principle has essentially two forms—weak and strong. The weak anthropic principle effectively says that, since we exist, then the nature of the universe we observe inevitably has to be such as to enable the development of intelligent carbon-based life in some regions. If it were not so then we would not be here to observe it! This is, essentially, a statement of the obvious and very few would argue with it. The strong anthropic principle goes further and states that the universe *must* have such a nature as to allow the development of life. The latter form is speculative and implies that the design of the universe and, in turn, the nature of a theory of everything are such as to ensure the existence of life. This may have attractions for theologians and, of course, can be related to the intentions of God if he/she exists, but this is trespassing into the realms of theology and belief which are not the direct concerns of this book.

The nature of a theory of everything, if it can be formulated, has still to be discovered. Much progress has been made towards possible structures but much more work is required. Not least is the need to achieve an understanding of the way in which the constants of nature emerge. Are there different universes in which different constants apply and are we in a special one in which the constants enable the development of life? Or, at the other extreme, is it possible that the values of the constants emerge from symmetry rather than design reasons as part and parcel of a final theory? Only time will tell.

11.4 Reductionism, Complexity, Determinism and Chaos

If a theory of everything is finally obtained does that mean that we shall then be in a position to *understand* everything? The answer to this question is uncertain. From the point of view of such a theory the material world and universe in which we live consist of fundamental particles, mostly protons, neutrons and electrons combined together in nuclei, atoms and molecules, interacting with each other through the strong, electromagnetic, weak and gravitational forces. The detailed behaviour of these entities is determined by the two key elements of the theory—quantum mechanics and general relativity. The reductionist approach, and this is the approach of physics, is then to attempt to understand the nature and behaviour of all forms of matter, at all levels of scale, in terms of the behaviour of such basic entities.

Obviously this becomes increasingly difficult as the complexity of the matter increases. However, we have already seen (Chapters 3 and 6) that many physical properties of matter *can* be understood in principle and in some detail in this way. This understanding was generally achieved by consideration of the *average* behaviour of large numbers of atoms and molecules. When we move on to consider chemical processes and the behaviour of very complicated molecules understanding is generally expressed in terms of chemical concepts such as bonds, valence, orbitals, acidity etc, but these concepts can all be explained in terms of interatomic interactions involving at a more fundamental level nuclei, electrons and the quantum treatment of their electromagnetic interactions. Moving into biology takes us into the extremely complicated realm of living entities whose patterns of behaviour are generally codified in a language quite different from that of physics. Here we encounter, for example, genes and DNA. But the nature of these entities can, again, be understood in the same terms as the simpler chemical structures. Moving into these different forms of matter involves a tremendous increase in complexity but not in the nature of the basic elements of this complexity. To put things bluntly: just as the adjective 'magnetic' is used to describe the nature of a material which can pick up iron filings so 'living' is the adjective given to the nature of a material which can reproduce itself. In both cases it is a question of how the basic components of the material are organized so as to account for their different natures. This is not to devalue the work of chemists and biologists. They are having to deal with extremely complex situations and just as physicists formulate their understanding of matter in concepts such as atoms, nuclei, forces, quantum mechanics, relativity etc so the chemists and biologists formulate their understanding in terms of appropriate concepts for their needs. It is simply that in the reductionist approach *their* concepts can generally be explained and understood in terms of those of physics.

But perhaps not always! For example, in considering brain behaviour, the concept of consciousness emerges—the fact that we, through our brains which consist of a complicated array of atoms and molcules, are concious of our existence. Can this conciousness be explained in physical terms? Some physicists believe that in due course we shall be able to do this; others have profound doubts. Again, only time will tell!

Continuing with this reductionist approach to understanding takes us on to the deterministic view of nature. In its extreme form, as discussed by Laplace in the 17th century, all particles of matter can be regarded as moving and behaving in ways specified by the 'laws of physics'—at that time, Newton's Laws of Motion. This meant that the universe could then be regarded as a gigantic machine whose future behaviour was completely determined by knowledge of its state at some past time. Taken to the extreme, this viewpoint removed the concept of free will: the future of the universe and all matter (inanimate or living) within it is determined from the outset!

However, there are two important developments since this idea was first propounded that mean that this perspective has to be abandoned. First there was the introduction of quantum mechanics and the realization that the precision of Newton's Laws of Motion did not hold when dealing with microscopic processes. The Heisenberg uncertainty relations (see sections 5.3 and 5.6) meant that it was impossible to give a completely precise description of the state of any particle or system. In addition, quantum mechanics only allowed a probabilistic description of future behaviour; we can, for example, only predict the *probability* of a radioactive nucleus decaying at a particular time.

Second, as mentioned in section 1.3 in connection with *chaotic* systems, there was the realization that even with a completely deterministic theory the future development of a system could be immensely sensitive to the precise nature of the initial conditions. Any small uncertainty in the initial conditions can multiply rapidly as time progresses making it quite impossible to predict future behaviour. Of course any uncertainty in the initial state of a system leads to some uncertainty in predicting its future. However, whilst with many sytems this uncertainty simply increases roughly in proportion to the time that passes, with chaotic systems the uncertainty grows *exponentially* with time. By exponential growth is meant one in which, if the uncertainty doubles after a time t, then it will double again after time $2t$, quadruple after time $3t$ and so on. This means that after time $10t$ the uncertainty will have increased not by a factor of ten but by a factor of $2^{10} = 1024$, and after time $20t$ by a factor around 10^6!

Chaotic systems vary from being quite simple in nature, for example a pendulum with a steel bob moving between two magnets, through to extremely complicated systems such as the weather. Such systems have been studied at many levels of scale, for example, the motion of atomic electrons, particles in accelerators, turbulence in fluids, cardiac rhythms, population biology and stellar motion, and may have played a part in the earliest moments of the big bang. Chaos can occur in systems governed by either classical or quantum mechanics. The former is completely deterministic and the latter deterministic as far as behaviour of the wavefunctions is concerned, although not in their physical meaning. Determinism is not the issue however; chaos results when the theory describing the system, deterministic or not, is *non-linear*. That is to say there is *feedback* in the theory such that the 'output' of its mathematical equations is not simply proportional to the 'input'. With the advent of powerful computers, the study of this sort of mathematics has been very intense over the last 25 years and it turns out that, in the midst of this complexity and unpredictability, certain general features in the behaviour of the system may emerge. For example, over time the system may keep going through approximately the same states (sensitively dependent on the initial conditions) and the mathematics of chaos theories helps us to understand these states, or, what are conventionally called *strange attractors*, but precise prediction is impossible.

In summary, for one reason or another, there is a great deal of uncertainty in the way in which many systems develop in time which is usually, but not always, linked to their complexity. It would seem that, generally speaking, the future is unpredictable in detailed terms both in principle and in practice. This is not to say that it will be at variance with our theories, simply that completely detailed manipulation of these theories and precise prediction are impossible.

11.5 Advancing Physics and Technology

We have seen that physics spans a wide spectrum of concepts and phenomena manifesting themselves at different levels of scale. At one extreme is elementary particle physics involving the study of extemely small entities at very high energies. Then, with decreasing energy and increasing size, we move through nuclear physics to atomic and molecular physics. Next comes condensed matter physics involving large numbers of atoms interacting together in the many different forms of matter. Finally, at the other extreme, is planetary and stellar physics and ultimately the physics of the universe.

To advance knowledge and understanding in these different areas involves experimenters, working in laboratories, and theoreticians who seek to understand experimental results in terms of current or, sometimes, new theories. Experimenters and theoreticians frequently work closely together, but not always. All of this work—research—is carried out in higher-education institutions, industry, government establishments and national and international laboratories. Its nature is frequently referred to as being either 'pure' or 'applied'. By 'pure' is meant research which is seeking to enhance fundamental knowledge and understanding—for example, in particle physics, astronomy and cosmology. By 'applied' is meant research which has technological aspects—for example, electronic devices such as silicon chips and even smaller (nanotechnology)—which, in due course, will lead to social and economic benefits.

However this nomenclature is somewhat misleading. To be sure some research is aimed solely at testing the predictions of a theory which, if confirmed, will then increase confidence in that theory. For example, experiments aimed at finding the Higgs boson (see section 9.5) fall into this category. Such research could, indeed, be described as pure in the sense that its aim is to increase knowledge and understanding of the physical world. However, in carrying out the relevant experiments, it may be necessary to devise and construct apparatus which is useful in other areas of technology; there is a technological 'spin-off'. Similarly, although applied research is focused on the development of materials, apparatus or processes satisfying a particular technological requirement it can frequently lead to new fundamental knowledge. Here might be cited the important contributions to the understanding of superconductors in the race to discover materials which exhibit superconductivity at high temperatures (see section 6.4). Basic physics and its technological ramifications advance together and are intimately related. Imagination and vision are required of physicists in both areas in order to effect this advance.

Here it must be recognized that there are frequently surprising outcomes in basic research and their implications can be unpredictable. Who would have predicted the profound implications of Einstein's '$E = mc^2$' for the development of nuclear weapons and nuclear energy? At a more mundane level, witness the development of nuclear magnetic resonance (NMR). This uses the fact that, since atomic nuclei are often magnetic (see section 8.2), when placed in a magnetic field they can exist in various energy states depending on their orientation with respect to the field. A study of the way in which they jump between these states, emitting or absorbing electromagnetic radiation, was researched for its own sake but then it was realized that not only can it give important information about the internal magnetic fields of a substance but also about the substance itself. It is used extensively in the study of materials and, most importantly, as a diagnostic tool in medicine where it appears as magnetic resonance imaging (MRI).

Of course funding is needed to support research. When the research has clear potential benefits for wealth creation then industry is willing to contribute and some industries are even prepared to support what is called 'blue-sky' research—research of an essentially pure nature but which *might* lead to important technological developments in the longer term. However when it comes to really pure or basic research then government financial support becomes essential. Inevitably as we probe further and further into physics such research becomes increasingly difficult and complicated, frequently requiring large teams of people working together. Above all, it becomes increasingly expensive. For example, higher-energy particle accelerators are needed; the increasing use of satellite observational systems to study the cosmos is proposed. Equipment at all levels of research becomes more sophisticated and it has become virtually impossible for any one nation, however wealthy, to provide the finance needed for its researchers. For this reason international collaboration is playing an increasingly important role in advancing research. This is particularly so in the fields of particle and nuclear physics (witness CERN: see section 9.1) and astronomy. Even this sharing of costs does not allow all developments that physicists would like to bring into being to take place and there are continual struggles to persuade governments to support research taking place in all countries.

Inevitably there has to be selectivity in the support of research at both the national and international level and the various bodies—political as well as scientific—concerned with such decision making are subject to a wide spectrum of arguments. As is to be expected, there are those who question the need, importance or relevance of seeking to understand the behaviour and nature of the physical world at its most fundamental level (e.g. quarks, the Higgs boson, the big bang, . . .). Replying to such doubters the basic argument is made that such research is a key element in the advance of human knowledge and in achieving a full understanding of nature and its universal truths. This is an aim and a purpose which is vital if we are to understand the nature of human existence.

11.6 What about Physicists?

Physicists are clearly motivated to advance physics and its associated technologies and they derive great pleasure from this activity. This pleasure can result from many aspects of their work—from the joy of solving a problem or developing a theory and so increasing understanding, from creating a new piece of apparatus or technique, from seeing the technological benefit of their work and, understandably, from external recognition of their standing and prestige—they are only human! The skills they develop as physicists can be described as problem solving, visionary, organizational, quantitative, communicative (with each other and, increasingly, with laypersons) and presentational. These latter skills arise since physics, indeed all science, advances most effectively when ideas, theories, techniques etc are shared and discussed both informally in seminars and lectures at institutions and at conferences together, of course, with the circulation and publication of papers as preprints (preliminary unpublished versions of papers), in journals and on the internet.

These qualities and skills of physicists mean that they are well equipped to undertake a variety of tasks and careers. A high proportion of those who train initially as physicists embark on very different successful careers in business, finance and administration. Not least important are their problem solving, quantitative and communication skills which are highly valued in many professional activities.

This is not to say that other qualitative human attributes, feelings and perceptions are suppressed. In particular they are very conscious that many technological developments resulting from the advance of physics have a downside to them—witness nuclear weapons, and the disposal of radioactive waste. Here many physicists have played an important role nationally and internationally, through organizations such as Pugwash (an international organization of scientists concerned with the control of developments in the area of nuclear energy and nuclear weapons), in trying to deal with such important moral and political issues. Neither is religious faith necessarily diminished by increasing confidence in the correctness of fundamental physical theories. The spectrum of beliefs of individual physicists, as in most other professions, spans atheism, humanism and agnosticism through to devout belief in the existence of God. They are much the same as everyone else. However the important thing that their training and experience as physicists brings to bear on their activities in life is a desire to understand what is happening in terms of more fundamental concepts, to seek a rational analysis of events and, not least, to use quantitative methods wherever appropriate. They recognize that physics is not the solution to everything but, nevertheless, see it is as the basic science for achieving understanding of the physical universe in which they live.

GLOSSARY

Mathematical Notation for Large and Small Numbers

Very large and very small numbers frequently appear in the text. The following concise notation is used.

10^n represents 1 followed by n zeros. In general, x^n means x multiplied by itself n times

$$\text{e.g. } 10^3 = 10 \times 10 \times 10 = 1000 \qquad 10^6 = 1,000,000$$
$$3.2 \times 10^4 = 32,000$$

10^{-n} represents 1 *divided by* (1 followed by n zeros). In general, x^{-n} means 1 divided by x multiplied by itself n times

$$\text{e.g. } 10^{-3} = 1/10^3 = 1/1000 = 0.001$$
$$10^{-6} = 1/10^6 = 1/1,000,000 = 0.000,001$$
$$3.2 \times 10^{-4} = 3.2/10^4 = 3.2/10,000 = 0.000,32.$$

Units

In scientific work an international system of units (*Systeme Internationale*—SI) is in general use.

The standard units and notation for length, time and mass are as follows:

Unit of length, 1 metre, denoted by 1 m
e.g. 3×10^3 m = 3000 metres

Unit of time, 1 second, denoted by 1 s
e.g. 2×10^{-2} s = 2/100 second

Unit of mass, 1 kilogram, denoted by 1 kg
e.g. 5.4×10^4 kg = 54,000 kilograms.

All other units of physical quantities can be expressed in terms of these basic units but, as illustrated below, are frequently given special names.

Other units which occur in the text are

Unit of speed, 1 metre per second, denoted by 1 m/s or $1\,\mathrm{m\,s^{-1}}$
    (see section 2.1)

Unit of acceleration, 1 metre per second per second, denoted by $1\,\mathrm{m/s^2}$ or $1\,\mathrm{m\,s^{-2}}$ (see section 2.2)

Unit of force, $1\,\mathrm{kg\,m\,s^{-2}} = 1$ newton, denoted by $1\,\mathrm{N}$
    (see section 2.2)

Unit of energy, $1\,\mathrm{kg\,m^2\,s^{-2}} = 1$ joule, denoted by $1\,\mathrm{J}$
    (see section 2.4).

(In atomic, nuclear and particle physics the electron volt is generally used as the unit of energy. This is defined in the next section.)

Unit of electric charge, 1 coulomb, denoted by $1\,\mathrm{C}$
    (see section 4.1)

Unit of electric potential difference, 1 volt, denoted by $1\,\mathrm{V}$
    (see section 4.2)

Unit of electric current, 1 ampere (or amp), denoted by $1\,\mathrm{A}$
    (see section 4.2)

Unit of electrical resistance, 1 ohm, denoted by $1\,\Omega$
    (see section 4.2)

Unit of power, 1 watt, denoted by $1\,\mathrm{W}$
    (see section 4.2)

Unit of rate of radioactive decay, 1 becquerel, denoted by $1\,\mathrm{Bq}$
    (see section 8.6).

Fundamental Physical Constants

These are constants which play a key role in determining the scale and nature of the physical universe.

Constant	Symbol	Value
Speed of light	c	$2.998 \times 10^8 \, \text{m s}^{-1}$
Planck's quantum constant	h	$6.626 \times 10^{-34} \, \text{J s}$
$h/2\pi$	$\hbar$	$1.055 \times 10^{-34} \, \text{J s}$
Gravitational constant	G	$6.673 \times 10^{-11} \, \text{N m}^2 \text{kg}^{-2}$
Mass of electron	m_e	$9.109 \times 10^{-31} \, \text{kg}$
of proton	m_p	$1.673 \times 10^{-27} \, \text{kg}$
of neutron	m_n	$1.675 \times 10^{-27} \, \text{kg}$
Proton charge	e	$1.602 \times 10^{-19} \, \text{C}$

The following is the unit of energy generally used in atomic, nuclear and elementary particle physics:

electronvolt	eV	$1.602 \times 10^{-19} \, \text{J}$
(see section 4.2)	$\text{keV} = 10^3 \, \text{eV}$	$1.602 \times 10^{-16} \, \text{J}$
	$\text{MeV} = 10^6 \, \text{eV}$	$1.602 \times 10^{-13} \, \text{J}$
	$\text{GeV} = 10^9 \, \text{eV}$	$1.602 \times 10^{-10} \, \text{J}$

Physical Terms

Brief definitions of technical terms used in the text.

absolute zero   The lowest possible temperature. The zero point of the Kelvin (or absolute) temperature scale.

acceleration   The rate at which the speed of a moving object changes.

accelerator   A linear or circular machine using electric and magnetic fields to accelerate charged particles to high energies.

alpha-decay   The decay of a radioactive nucleus involving the emission of a helium nucleus (alpha-particle).

ampere   The SI unit of electric current.

amplitude   A measure of the extremes of motion of, for example, a wave or oscillating system.

angular momentum   A measure of the vigour with which a body rotates.

anthropic principle   The universe we observe is the way it is because if it were different we would not exist.

antiparticle   A particle which has the same mass and spin as the particle to which it corresponds but with other properties (e.g. charge) reversed.

asymptotic freedom   The property of the force between quarks referring to the fact that when quarks are very close to one another they no longer experience the force and are therefore effectively 'free'.

atom   The basic unit of ordinary matter consisting of a nucleus with a positive electric charge together with a number of electrons whose total negative charge has the same magnitude as the positive charge.

atomic number   The total number (denoted by Z) of protons in a nucleus. It is equal to the number of electrons in the corresponding neutral atom.

band   A group of closely spaced energy levels available to electrons in a metal or semiconductor.

baryon Those elementary particles having half-integer spin which experience the strong interaction.

baryon number The number +1 or −1 given to baryons and antibaryons respectively. All other particles have baryon number 0 and in any elementary particle process the total baryon number is additively conserved.

becquerel The unit used for specifying the rate of radioactive decay.

beta-decay The decay of a radioactive nucleus involving the emission of either an electron and an antineutrino or a positron and a neutrino.

big bang The massive pointlike explosion initiating the expanding universe.

binding energy The energy needed to completely disintegrate a nucleus into its separate neutrons and protons.

black hole A body which is so dense and has such a strong gravitational field that light cannot escape from it.

Bohr model The model of an atom in which electrons are restricted to circulate about the atomic nucleus in orbits such that the electron angular momentum is an integer multiple of $\hbar$.

boson A particle whose spin is either zero or a whole number (0, 1, 2, . . .).

bottom A 'flavour' allocated to quarks which is conserved additively in strong and electromagnetic processes.

brown dwarf The name given to cool stars whose masses are too low to allow nuclear fusion to take place within them.

bubble chamber A device containing liquid near to boiling which reveals the tracks of charged particles passing through it by the string of bubbles they create.

centripetal force The central inward force holding an object in rotational motion.

chain reaction An escalating series of fission processes.

chaos The random and unpredictable behaviour of non-linear dynamical systems such as the weather.

charm   A 'flavour' allocated to quarks which is conserved additively in strong and electromagnetic interactions.

chromosphere   A layer of rarefied gases near the surface of the sun.

collective model   A model of the atomic nucleus in which the component nucleons oscillate or rotate *en masse.*

colour   A charge (red, yellow or blue) allocated to quarks analogous to electric charge and which measures the strength of the inter-quark force.

compound nucleus   The intermediate short-lived state in a nuclear reaction formed when a bombarding particle is captured briefly by a target nucleus.

conduction   The transfer of heat or electricity through a material object without movement of the material.

conduction band   A band of energy levels in a metal in which electrons are free to move through the metal.

confinement   The property of the force between quarks referring to the fact that as two quarks separate from each other the force of attraction between them continues to increase so that, whatever energy is provided, they are unable to escape from each other and exist in isolation.

convection   The transfer of heat in a liquid or a gas by bulk movement of matter.

corona   Dust particles at very high temperature in the surface of the sun.

cosmic microwave background radiation   Electromagnetic radiation, mostly in the microwave region, which pervades the whole of space and is a relic of the big bang.

cosmic rays   High-energy particles reaching the earth's atmosphere from outer space.

cosmological constant   A constant introduced by Einstein into the general theory of relativity to enable the theory to accommodate the idea of a 'static' rather than an 'expanding' universe.

cosmological principle   The large-scale uniformity of the visible universe.

coulomb   The SI unit of electric charge.

counter   A device for detecting and counting charged particles.

covalent force   An interatomic force resulting from some sharing of electrons between the two atoms.

critical density   The density of matter in the universe such that the universe would just expand to infinity.

critical mass   The mass of fissionable material needed to sustain a chain reaction.

dark matter   Matter which exists in the universe but which is not directly detectable.

de Broglie wavelength   The wavelength ($= h/p$) of the quantum mechanical wave associated with a particle having momentum p.

density   The mass per unit volume of a substance.

detector   A device for detecting elementary particles and ions.

deuteron   The lightest nucleus consisting of a bound state of a neutron and a proton.

diamagnetism   The magnetism induced in a substance by an external magnetic field which is opposite in direction to the inducing field.

diffraction   The modification of light or other wave motion in passing, for example, through a narrow slit.

diode   An electronic device that allows the flow of electric current in one direction only.

Dirac Equation   A relativistic wave equation which describes the quantum behaviour of particles, such as the electron, with half-integer spin.

dispersion   The spreading of light into its different colours to form a spectrum by, for example, refraction.

domain   A small region of a ferromagnetic material in which all the atoms have their magnetism aligned.

donor An impurity introduced into a semiconductor, some of whose electrons contribute to the semiconductor's conductivity.

doping The introduction of an impurity into a semiconductor in order to increase its electrical conductivity.

Doppler effect The change in the apparent frequency of a wave when there is relative motion between the source and the observer.

ecliptic The plane in which planetary orbits are roughly contained.

electric charge The property (positive or negative) of a particle which determines the size and nature (repulsive or attractive) of the force it experiences in an electric field.

electric current The flow of electric charge through a conductor.

electric field The region around an electric charge in which another electric charge experiences a force.

electromagnetic The adjective used to describe any phenomenon (e.g. a wave) involving electric and magnetic fields.

electromagnetic induction The inducing of an electric current in a wire or coil by its motion relative to a magnetic field.

electron A light elementary particle carrying negative electric charge and having spin $\frac{1}{2}$.

electroweak theory A unified theory of elementary particle interactions which accounts for both electromagnetic and weak interaction processes.

entropy A measure of the amount of disorder in a system.

exclusion principle The principle due to Pauli that no two identical half-integer spin particles (e.g. electrons) can occupy the same quantum state.

fermion An elementary particle having half-integer spin.

ferromagnetism A form of magnetism in substances such as iron which can be present permanently without the presence of an external magnetic field.

Feynman diagram A diagram showing the creation and annihilation of particles during an elementary particle interaction process.

field A region of influence (due, for example, to an electric charge—an electric field) existing in space and time.

fission The splitting of atomic nuclei with the release of large amounts of energy.

flavour A quality such as 'strangeness', 'charm' etc assigned to quarks and elementary particles.

force That influence which, if applied, causes a body to change its state of motion.

frame of reference An arbitrary set of axes to which the position or motion of something in space and time can be referred.

frequency The number of complete waves passing a point, or the number of complete oscillations of an oscillating system, in a second.

friction Resistance to relative motion between two bodies.

fusion The combining of light atomic nuclei to form heavier nuclei resulting in the release of a large amount of energy.

galaxy An independent system of many stars together with associated dust and gases.

gamma-decay The decay of a nucleus with the emission of very short-wavelength electromagnetic radiation (gamma-rays).

gauge boson One of a group of integer-spin particles (such as the photon) transferred between elementary particles to propagate the strong, electromagnetic, weak and gravitational interactions.

general relativity The theory developed by Einstein based on the postulate that the laws of physics should be the same in all frames of reference whatever their motion and which describes gravity as a geometric property of four-dimensional space–time.

geodesic The shortest path between two points in space–time.

gluon The gauge boson (with spin 1) propagating the strong interaction between quarks.

glueball   A collection of gluons bound together for a short period.

grand unification theory   A unified theory encapsulating the strong, electromagnetic and weak interactions.

gravitational mass   The attribute of an object which determines the force it will experience in a gravitational field.

graviton   The quantum unit of gravitational radiation.

gravity   The universal force of attraction which exists between all forms of matter.

hadron   An elementary particle which experiences the strong interaction as well as the weak and the electromagnetic interaction.

heat   The form of energy due to the motion of atoms and molecules in a substance.

Higgs boson   A so-far undetected heavy particle which is a manifestation of the 'Higgs field', which is postulated to account for the masses of elementary particles.

Hubble's law   A statement of the fact that the speed of recession of distant galaxies is proportional to their distance from us.

inertial frame of reference   A frame of reference in which Newton's laws hold exactly.

inertial mass   The property of an object which determines the extent to which its state of motion is changed when an external force is applied.

inflation The hypothesized brief, but gigantic, expansion of the universe immediately after the big bang.

insulator   A substance which is a bad conductor of electricity or heat.

integrated circuit   An electronic circuit embedded in or on a slice of semiconductor material.

interference   The effect of two waves meeting. The different combinations of their crests and troughs leads to the formation of an 'interference pattern'.

internal energy   The energy possessed by a substance by virtue of the motion of its atoms and molecules.

ion   An atom which has lost or gained one or more electrons so that it has a resultant net electric charge.

ionic force   The electric force between two ions.

ionization   The process of converting an atom into an ion.

isotope   Any one of a group of nuclei which have the same atomic number (i.e. same number of protons) but different mass number (i.e. different numbers of neutrons).

joule   The SI unit of work or energy equal to the work done by a force of 1 N in moving through a distance of 1 m.

kelvin   The SI unit of temperature defined by the Kelvin (or absolute) temperature scale.

kinetic energy   The energy an object possesses by virtue of its motion.

kinetic theory   The theory, particularly of gases, based on the motion and collisions of individual atoms and molecules.

Klein–Gordon Equation   A relativistic wave equation which describes the quantum behaviour of particles, such as the pion, with integer spin.

laser   An acronym for light amplification by stimulated emission of radiation.

LED   Light emitting diode. A diode that emits light when an electric current passes through it used for the display of, for example, numbers in an electronic calculator.

lepton   One of a group of six elementary particles of spin $\frac{1}{2}$ which do not experience the strong interaction.

lepton number   A number (±1) allocated to all leptons (it is zero for all other particles) which is additively conserved in any elementary particle process.

light year   The distance that light travels in one year (equal to 9.46×10^{15} m).

liquid drop model A model of the atomic nucleus in which certain nuclear properties can be understood if the nucleus is treated as a drop of liquid.

Lorentz invariance The invariance of the laws of physics in special relativity under the Lorentz transformation in which the space–time coordinates in one inertial frame of reference are transformed into those measured in another inertial frame moving relative to it with uniform speed.

magic number Nuclei which contain these numbers of neutrons or protons are particularly stable. They correspond to the filling of particular shells in such nuclei.

magnetic field The region around, for example, a magnet in which another magnet would experience a magnetic force.

mass number The total number (denoted by A) of neutrons and protons in a nucleus, which is roughly proportional to the nuclear mass.

mean life The average life of a radioactive nucleus (or, for example, an excited atom) before it decays.

meson A hadron which is also a boson (i.e. has integer spin).

molecule The smallest unit of a substance consisting of two or more atoms and which has the chemical properties of that substance.

momentum A vector quantity equal to the product of the mass and the velocity of a moving object.

monopole A hypothetical free standing magnetic north or south pole.

muon A charged lepton having mass around 207 times that of the electron.

neutrino The name given to a neutral lepton.

neutron A neutral baryon which, together with the proton, is a component entity of an atomic nucleus.

neutron star A very small star that has collapsed under gravity and which consists essentially of neutrons.

newton   The SI unit of force equal to that force which will give an acceleration of 1 m s^{-2} to a mass of 1 kg.

nuclear reactor   An apparatus in which self-sustaining fission reactions occur with the release of large amounts of energy.

nucleon   The collective name for neutrons and protons.

nucleus   The central core of an atom consisting of neutrons and protons held together by the strong nuclear force and having a positive electric charge.

ohm   The SI unit of electrical resistance.

Ohm's law   The relation between the current in a wire, the wire's resistance and the applied potential difference.

pair production   The simultaneous production of a charged particle and its (oppositely charged) antiparticle—usually an electron and a positron—by electromagnetic radiation.

paramagnetism   The weak magnetism which is produced when the individual atomic 'magnets' in a substance are partially 'lined up' by an applied magnetic field.

parity   The property of a wave function in quantum theory which specifies whether it remains the same (even parity) or changes sign (odd parity) when subject to mirror reflection (i.e. the signs of all spatial coordinates are reversed).

perihelion   The point in the orbit of a planet which is nearest the sun.

photoconductor   A substance whose electrical conductivity is influenced by exposure to light.

photoelectric effect   The emission of electrons from the surface of a metal when exposed to electromagnetic radiation (e.g. light).

photon   A quantum 'package' of electromagnetic radiation carrying energy equal to the product of Planck's quantum constant, h, and the frequency of the radiation.

photosphere   The outer region of a star's surface, from which light is emitted.

pion   An unstable meson which can have positive, negative or zero charge, whose exchange between nucleons contributes significantly to the nuclear force.

Planck's constant   A fundamental constant denoted by h, having the dimensions of angular momentum. $h/2\pi$ is always denoted by $\hbar$.

positron   The antiparticle of an electron.

potential difference   An expression (measured in volts) for the difference in potential energy of a unit charge (one coulomb) between two points in an electric field.

potential energy   The energy possessed by a piece of matter by virtue of its position (e.g height above the ground) or state (e.g compressed or stretched).

pressure   The force (measured in newtons) exerted per square metre by a gas on the surface of its container.

proton   A spin $\frac{1}{2}$ hadron carrying unit positive charge which, together with the neutron, is a basic component of an atomic nucleus.

pulsar   A rapidly rotating neutron star emitting regular pulses of electromagnetic radiation.

quantum chromodynamics   The quantum theory of the interaction between quarks and gluons.

quantum electrodynamics   The quantum theory of the interaction between elementary particles and electromagnetic radiation.

quantum mechanics   That branch of physics dealing with the mechanics and physical behaviour of atoms and other particles of matter.

quantum number   Integer or half-integer numbers which define the quantum state of a physical system.

quark   One of a group of six basic particles carrying fractional electric charge which are the fundamental constituents of all strongly interacting elementary particles.

quasar   A very bright quasi-stellar celestial object believed to result from the gravitational collapse of the central region of a galaxy.

radioactivity   The spontaneous emission of alpha-, beta- or gamma-rays from an unstable nucleus.

red giant   A star which has undergone massive expansion and cooling.

red shift   The shift towards the red end of the spectrum of the light emitted by receding stellar objects due to the Doppler effect.

refraction   The bending of a light beam as it passes from one transparent medium to another.

rest mass   In relativity theory, the mass of a body when it is at rest relative to an observer.

Schrödinger Equation   A mathematical equation in quantum mechanics which determines the wavefunction and associated physical quantities of a system.

semiconductor   A substance such as silicon whose small electrical conductivity is increased by raising its temperature or by adding impurities (dopants).

semileptonic decay   A particle decay process involving both hadrons and leptons.

shell model   A model of the atomic nucleus in which the individual nucleons move in quantum states in a potential well, forming 'shells' of nucleons.

simple harmonic motion   The oscillatory motion of, for example, a simple pendulum in which the frequency of oscillation is independent of the amplitude.

SI units   An international system (Systeme Internationale) of units used in physics.

space–time   The four-dimensional (three space and one time) reference system used in relativity theory.

special relativity   The theory developed by Einstein based on the postulates that the laws of physics should be the same in all frames of reference in uniform motion relative to each other and that the speed of light is the same in all such frames.

spectrum The different frequencies of the electromagnetic radiation emitted by a physical system such as an atom or a nucleus.

spin The intrinsic angular momentum of a particle which can be an integer or half-integer multiple of $\hbar$.

spontaneous emission The spontaneous emission of radiation by, for example, an excited atom.

spontaneous symmetry breaking The exhibition of differences between, for example, weak and electromagnetic interactions at low energies which do not exist at high energies when both are described by a unified theory.

standing wave A wave motion in or on a confined medium, e.g. a wave on a violin string.

stimulated emission The emission of radiation by, for example, an excited atom caused by incident electromgnetic radiation. It is the basic process responsible for the operation of a laser.

strangeness A 'flavour' allocated to strongly interacting particles (hadrons) which is conserved additively in all strong interaction processes.

strong interaction The strongest of the four basic interactions (strong, electromagnetic, weak and gravitational) which, at the fundamental level, acts between quarks and gluons. It is responsible for the nuclear force.

superconductivity The disappearance of electrical resistivity of certain materials at relatively low temperatures.

superfluidity The state of a liquid (e.g. liquid helium) below a certain critical temperature when it is able to flow and conduct heat without experiencing any resistance.

supernova A catastrophic stellar explosion in which the outer layers of the star are blown into space.

superstring theory A theory in which basic elementary particles are represented by miniscule vibrating strings.

supersymmetry   A theory of elementary particle interactions in which the basic spin-$\frac{1}{2}$ particles such as quarks and electrons have partners with integer spins and, correspondingly, integer-spin gauge bosons have partners with spin $\frac{1}{2}$.

synchrotron radiation   The electromagnetic radiation emitted by high-energy accelerated charged particles.

temperature   The degree of hotness or coldness of a body as measured on a temperature scale such as Celsius or Kelvin (or absolute). On the latter scale the temperature is proportional to the internal kinetic energy of the body's component molecules.

thermodynamics   That branch of physics which deals with the relationship between heat, energy and mechanical action.

time dilation   The effect in relativity theory of time 'slowing down' due to speed of travel relative to an observer or due to gravity.

top   A 'flavour' allocated to quarks which is conserved additively in strong and electromagnetic processes.

torque   A twisting force which changes the rotational motion of a body.

transistor   An electronic amplification device (a triode) consisting of semiconductors having common physical boundaries.

travelling wave   A wave motion in which the wave shape travels continuously in a particular direction.

uncertainty principle   A prinicple in quantum mechanics which asserts that there is a limit on the accuracy with which simultaneous measurements of certain pairs of physical quantities (e.g. position and momentum) of a particle or system can be made.

van der Waals force   A weak attractive force between neutral atoms arising from distortion of their electron clouds.

vector   A physical quantity having both magnitude and direction (e.g. a force).

velocity   A vector quantity specifying speed and direction of travel.

volt   The SI unit of electric potential difference.

watt   The SI unit of power.

wavefunction   The mathematical function in quantum mechanics which describes the physical state of a system and whose intensity, for example, specifies the probability of finding a particle at a particular point.

wavelength   The distance between successive peaks or troughs of a wave motion.

wave–particle duality   The concept in quantum mechanics that particles can behave like waves and waves like particles.

W-boson   A gauge boson, with positive or negative electric charge, responsible for propagating the weak interaction.

weak interaction   The interaction in elementary particle physics responsible for weak processes such as beta-decay.

weight   The gravitational force by which a body is attracted to a celestial body (e.g. the earth).

white dwarf   The final state of certain stars which have no more fuel for nuclear fusion and in which complete gravitational collapse is prevented by the repulsion between electrons due to the exclusion principle.

wire chamber   A device for detecting charged particles consisting of charged wires immersed in an appropriate gas.

work   The energy imparted to a body when it is moved by a force. It is equal to the product of the force and the distance the body is moved by the force.

x-rays   Very high-energy electromagnetic radiation.

Z-boson   An electrically neutral gauge boson responsible for propagating the weak interaction.

zero-point energy   The lowest energy a confined particle can have in a quantum system. It is not zero because of the limits imposed by the uncertainty relation.

INDEX

Milton Keynes UK
Ingram Content Group UK Ltd.
UKHW022051141024
449569UK00031B/1584